MW01245273

OUR STORY

Personal Experiences of Submarine Service and Other Memories

by

Rick Palumbo

ISBN 979-8-9872572-0-3

Cover design and interior formatting
by Mandy Haynes
@ threedogswritepress

This book is dedicated to all those affected by ALS (Amyotrophic Lateral Sclerosis), commonly known as motor neuron disease or Lou Gehrig's disease in the United States. A portion of the sales of *Our Story* will be donated to the ALS Association in honor of those afflicted by this disease and their family and friends.

Table of Contents

Introduction

This story details the experiences of Dan and his ultimate decision to join the Navy, volunteer to serve on board a submarine, and his experiences aboard that submarine. The story is a personal perspective and memories of actual events and people, both in and out of military service, with a focus on events involving his submarine service. The story begins with college graduation and ends at the end of Dan's six-year commitment. Many military service and civilian-related anecdotes are recounted.

Dan's decision to follow that path was difficult. In hindsight, it gave him direction and purpose, but it was only after reaching a point of persistent depression and a feeling that his life was at a dead end that he took a chance on himself through submarine service.

The story begins at Dan's college graduation a couple years prior to him even considering the military. This third-person recounting by his good friend, Ted, describes the two years that led to Dan's need to make a drastic change. That change or milestone event, as he'd define it, was the decision to commit to a six-year stint in the Navy. Ted describes Dan's personal story and path to a decision to join the Navy. Ted's perspective explains not only why the military life was chosen, but how Dan arrived at the decision to consider the military and submarines specifically.

Dan himself provides a first-person account of his experiences in the Navy and aboard a submarine. He provides stories and insight into his experiences. It is something most understandably cannot relate to. Often, people seem to have difficulty deciding what questions to ask. The limited understanding they have is from suppositions made from personal deductions or exaggerated stories in movies that don't reflect real life. However, part of any submariner's story is the experiences from training and skills obtained in the Navy prior to even stepping aboard a sub - like events in boot camp, submarine school and other job training. And even given all that, no one volunteers for submarines, the Navy or the military, in general, in a vacuum.

One's life experiences affects every perception, decision and long term memory. It drive's judgement and affects personality. Therefore, it is impossible to avoid the complications and experiences that everyday life presents when telling the stories of the military experience. They are intertwined to be sure.

Dan's attempt to meet his own expectations as well as those he perceives of others takes a toll on him. The experiences and decisions he develops and cultivates in this period of time determine his direction for the rest of his life. Some of the relationships he develops are experiences Dan maintains as strong memories. The outcome of a few personal relationships are detailed and impactful in defining his future, but a significant portion of this story focuses on the Navy and submarine experience – honesty with doses of hopeful dry humor.

In a very off-hand way, some veterans outside of the submarine service casually describe submarine service as a specialty within the military, but it's more than that.

Submarine service is unique in that it's 100% voluntary. No one in the Navy is assigned that duty. Each service member on board volunteered for the long periods of isolation that Dan experienced.

His story provides insight and realism so civilians might somewhat relate to or at least increase their understanding to some extent.

Imagine a life defined by its past,
Or an abandoned and wandering ship absent a sail on its mast.
If hope can be the energy illuminating the fork in the road,
Which path then, for which we'll be known,
and what friends to lighten our load?

Chapter 1

College is Finally Over!

But First, September 1989

Activity on the ship was hurried and hectic. Anything and everything must be secured to quickly get to sea and ride-out the hurricane. Certainly everyone on board was incredibly worried for any family they felt they were abandoning, but our focus had to be on preparing the ship for sea. For most, it was unknown how much time they had until the effects of the storm would be felt or how long they'd have to endure the storm once it started. A mystery to most as well was what a hurricane feels like, especially when immersed in it at sea. We safely left the harbor, as did all the other ships and submarines and entered open ocean with Hurricane Hugo soon to envelope us.

We were already 'rocking and rolling' or 'bobbing and rolling', as my fellow sailors would say. That evening, I was assigned to be a lookout on the aft part of the ship. Thankfully, there was still spotty daylight, but with quickly deteriorating conditions. My preparation included instruction to alert the bridge of any sightings and include their relative position to our ship as well as if the sighting's direction of travel could be determined. Hopefully, any other ships in the area would have their lights on since vision was minimal at best. I was in a sense, packaged, in a heavy one-piece waterproof outfit with a hood, sound-powered headphones, binoculars, and a harness with a lanyard.

Upon donning this equipment, I left the relative comfort of the ship that was now feeling the considerable effects of the hurricane as I walked outside through a blinding rainstorm to relieve the lookout whose place I was taking. I secured myself to the ship with the lanyard attached to my harness so as to prevent being blown overboard. The person I relieved double checked that I was secured, then left for the interior spaces. I then began looking for anything.

Waves made sailing very rough and rain made visibility minimal. We were being tossed about unpredictably. Impressive waves with white caps slammed into the side of the ship, washed over the deck and sprayed over me. I had to grab any available handholds to prevent being knocked to the deck.

Seeing past the edge of the ship was soon impossible. My only communications with the bridge was for them to verify I was still in place. The wind driven rain was almost painful, beating me physically. I was thankful for the thick waterproof clothing in helping to absorb the punishment. After way too long, it seemed to me (approximately three hours), my relief came and I returned to the relative safety inside the ship. I never did note any contacts and was exhausted from fighting the storm.

I'm sure we weren't, but it seemed we were alone out there. Sleep was minimal and with constant interruption as the hurricane had its way with the ship. I thought of the subs that were probably safely at considerable depth. Sailing experience has taught me how to secure myself in my bunk so the tossing and turning of the ship would not throw me to the deck.

A new day didn't bring relief from the storm. My normal duties as a quality control inspector were not needed, so I was assigned to take a shift on the bridge of the ship. At least this time, I won't be exposed to the pounding of the rain. It allowed me to see and experience a hurricane very few are permitted. I was responsible for the ship's speed (sort of). The Officer of the Deck (OOD) would bark an order as to the speed he wanted. I relayed that to the engineering spaces who would make that happen and subsequently respond to me when the order was accomplished. I then informed the OOD.

Communication on the bridge was difficult at best. Because of the storm, an increased focus and a loud voice was required. Everyone “talked” to each other by yelling. The storm was incredibly loud. Waves hitting the hull with unpredictable blunt force and unpredictable periodicity produced a loud hammering sound. The view out the front of the ship and over the bow was of enormous waves with intense whitecaps producing spray seemingly as high as the ship was tall. The waves would hammer the ship with amazing power, as they broke over the bow, pounding the windows and flooding the exposed surfaces of the ship. The waves loudly slapped the windows, causing my head to instinctively snap back and my eyes blink rapidly. I wasn't confident the windows could withstand the punishment.

Our ability to steer was limited. We all constantly looked back and forth to help support anyone who didn't properly counter the ship's movement and stumble or fall to the deck. Then suddenly the ship dove into a cavernous space between waves, darkness surrounded the ship as we disappeared in this void, only to have the ship rise again and have the process repeat itself. This repetition was exhausting. Just maintaining a standing position was difficult. There was nothing to do, but anything necessary to survive. Had not everyone been so focused on their jobs, fear would probably be the only recourse.

The hurricane pushed itself ashore at Charleston and then

swallowed the state. It, of course, caused impressive damage to South Carolina and beyond. We returned to port, only to sit off-shore in view of the coastline and wait. Time was needed to get soundings of the ocean and river. The Navy had to be sure ships wouldn't run aground due to the hurricane changing the ocean's and river's bottom and depth. Standing topside, one could observe a beach community leveled, every tree snapped or toppled and entire houses from the destroyed neighborhoods floating past us in the ocean.

I and others were left to wonder if our families made it. There was no way to know where they were, how they fared or if the worst happened.

Days passed.

December, 1981

I feel like I know Dan as well as anyone, including his family. We've been best friends since second grade. We sat next to each other whenever the teacher allowed it throughout grade school. As next-door neighbors, we were always together every day after school and the entire summer year-after-year. Looking back, I believe we started to drift apart in high school. We had many of the same friends, but time together was less. It wasn't a conscious decision. There was never a falling-out. We just slowly and unknowingly were apparently moving on different paths–an observation that's clear mostly in retrospect. We graduated from college the same year; although from different institutions and me in May and Dan in December.

Following the last words of the final graduation speech, a loud yell filled the venue and almost everyone's caps filled the air as the visitors clapped. His cap remained in place. While most graduates were overwhelmed with joy, his response was muted. For him, this all might be just ceremonial. His parents were among the visitors clapping and smiling. Their pride was truly evident, but they had no idea why his pride in himself was absent.

It was December, so this graduating class was smaller, hardly filling half of the basketball arena. Visitors all had some distance to travel, as this was a small rural university. Weather cooperated. It was cold, but expectedly so, with hardly a breeze to effect a wind chill. The sun shone brightly past a smattering of high, thin clouds, yielding a bright blue sky. Most needed only light jackets since wind was minimal to absent. No snow had fallen yet this season. Therefore, walkways were dry and easily traversed.

Dan and his parents left the venue, but not to celebrate with a meal or a gathering of additional family and friends. Instead, he hurriedly ushered his parents from the venue to their car, providing them directions as they all drove to the chemistry building, Hughes Hall, where he hoped to see his chemistry teacher. The short car ride was with him alone in the back seat, his gown still draped around him and cap on the seat next to him. When his mother inquired as to the reason for coming here, Dan said he'd explain later. Dan was obviously nervous. They didn't converse, but it took only minutes to reach their destination. The chemistry building was very familiar to Dan. He had multiple classes there every year of the last four years of college.

To be sure, Dan should have used his time better over the last four years. Admittedly, his dislike of his college major was an unfortunate problem, but he was probably not ready to be on his own. Dan was immature. He didn't exhibit much discipline or apply himself nearly enough. This wasn't the first class Dan had concerns about passing. The last four years are full of good memories though. He joined a fraternity and made many good friends. While Dan certainly focused more on having fun than showing responsibility towards a degree, it is unfair to say what time he did spend studying was a waste. He wasn't failing. If his chemistry grade prevents him from graduating, he'll surely make it next semester. And although a student's GPA and accumulated grades are the only determination of a degree, there is certainly more to college. He's shown that in his participation in athletics and strong work ethic outside of the classroom. In an unusual display of good decision-making, Dan took an economics class over one summer at a local college near home to make himself more attractive to potential employers. But, as was his routine, he didn't apply himself and his final grade showed it. So, in the final analysis, his impressive effort in this chemistry class was really too little, too late.

Now arriving at the chemistry building, Dan leapt from the car, leaving his graduation cap on the seat and directing his father to sit tight, telling him, "I won't be long."

His graduation gown waved like a cape as Dan ran to the entrance of the building and jogged down a short hall to steps leading to the basement where he hurried down another hall to his professor's office. Dan knew this path well. The building was empty except for a smattering of university staff. He had been here often over the previous four years, but especially throughout the previous semester requesting help and for answers to questions, so I assume they knew each other at some level. Luckily, his teacher was there and the door propped open. This hopeful graduate knocked softly and politely approached the large and opposing wooden desk, cluttered with stacks of paperwork. As it was obvious he had just left the graduation ceremony, Dan calmly, but bluntly asked if he really graduated. This class was a requirement and he was very nervous. Without even a pause to look-up his actual grade, the professor responded, "Don't worry, I don't fail seniors."

A rising joy and small pulse of adrenaline filled Dan's body.

"Thank you," he said, then Dan turned and reversed his path to exit the building at an even quicker pace to the car without waiting for any additional words from the professor. In the years since, Dan voiced curiosity of what his true grade would have been.

Maybe he really did pass, Dan sometimes considered. He worked and studied hard for this class. Sometimes it just doesn't seem to matter or seem fair, he believed. He attended study groups, personally sought the professor's help and put in long hours preparing for classes and tests. It wasn't his grade point average that worried him. It also wasn't good but was good enough. This class was a requirement for his degree, so passing it was imperative, but this class was quite difficult. Many students dropped out, leaving less than half that actually started it to take its final exam. The four-and a-half-year trek through college finally was at an end.

Dan jumped into the back seat of the car his father kept running. A genuine smile and sense of obvious relief on his face greeted his parents, finally. He provided an explanation purposely devoid of detail to his parents of this strange side trip. They remained proud and stated as much. Dan would never talk of that short trip again to his parents, but it's likely they surmised its purpose.

Stopping only to get the last of his things, he and his parents left campus, never to return. The time spent getting this degree was not productive. The degree Dan earned was not as exciting or interesting as initially thought during that high school visit seemingly so long ago. He learned that early in his sophomore year, but could not afford, financially, to change degrees. He got through it.

I suppose Dan deserves credit, but it should be tempered. The required classes were quite difficult, especially when one is disinterested and immature. At the time, his degree had the reputation as being the most difficult offered by the university. Dan literally hated the degree he was awarded and was saddened at the thought of continuing to pursue it in the workforce, since it was a degree very specific to the paper industry. The long ride home with his parents was quiet. Dan wasn't in the mood to talk. He thought about the last four years, his grades and his experience with that last chemistry class.

Dan remembers never really feeling like his major was his choice. He had often thought about doing something in the medical field, but he wasn't *doctor material* and wasn't aware of other options except nursing. In those days, there existed a bias that men became doctors and women nurses. He feels he has accomplished nothing since high school graduation. Regret can be overwhelming. I often wondered about his condition at this point in his life.

Upon graduating high school, there seemed to be so much promise. It certainly was long lost now. Dan now seemed to express something more than just self-doubt. I think body language suggested an expression of very little self-confidence and a belief he would never meet expectations, but I doubt if Dan's family ever talked of such topics. It's a difficult discussion for anyone.

Before long, they would arrive home to a small one and a half story red brick house with a fenced yard in what most might describe as a typical suburban neighborhood in those days. Yards were small. It was a mostly white middle-class environment.

It offered a comfortable life.

Her life was similar…

The small two story New England style house in this northern town sat on a small, fenced lot with a detached garage, but with an above ground pool that they and their neighbors used during the summer. The house and garage had clap-board painted white. They were a relatively large catholic family with five kids and two medium-sized dogs. The neighborhood was filled with similar houses containing mostly white, middle-class families. The old age of this suburb was revealed by the very large mature mostly oak trees dotting the length of the street.

Her family made do with a single bathroom, but with an additional shower hastily added in the basement for anyone not willing to wait. Privacy among the siblings was limited. In recounting her childhood, nothing particularly exciting ever happened, but she still had fun times she'd recount. Grade school was close enough to enable walking. High school required driving.

She was pretty, blonde, average height, and well adjusted. Her hair extended to a length just below her shoulders. She occasionally created curls, but most of the time kept it relatively straight with a slight waviness. Her friends were similar to her in their blue-collar background. She wasn't popular, as a 'high-schooler' might define it but was also neither a nerd nor a jock. Just what her friends described as a nice person. She was the oldest of the five kids and often made to baby-sit her two youngest siblings.

School was approaching the winter break. The cold

weather had set-in. Snow has already made an appearance. Each day was a repeat of grey clouds with varying chances of frozen precipitation. A small amount of snow accumulation has come and gone. A break from classes was welcomed; although no such break existed for her at work.

At this point in her life, it was December and she was mid-way through her junior year in high school. Being the oldest sibling, she was allowed well-monitored independence. She got away with some things, but her time was closely monitored due to her being female and the eldest child. Her parents were conservative and new at raising teenagers. She has been a live-in babysitter for her much younger siblings – not a decision she had a choice to turn down. This was one reason she sought her independence.

She had decided to pursue technical school for the final two years of high school in the hopes of becoming a hair stylist. College wasn't often discussed and not her goal. Her singular goal was to become independent. Styling hair was a means to that end. She didn't waste time on lofty goals, like college seemed to be. She was raised that way. Her parents didn't talk of college. If she wanted it, she'd have a lot of leg-work to do and have to figure out on her own how to pay for it, but her impressive work ethic did allow her to maintain a job at a local fast-food establishment while attending the two-year trade school.

She didn't see herself as smart but was wrong in that assessment. Life was a boring daily repeat of styling classes during the day followed by work in the evening. At this point, half of her junior year was done. Her counting the months was a way to keep her mind on her goal.

His days after college…

Dan and I were quite good friends. We have been for years, beginning in second grade. Our time as friends involved many a personal discussion. Sometimes we just complained at our present condition and sometimes just listened as the other vented. We both harassed and supported each other throughout our years of friendship. As I speak today, it's been a number of years since Dan and I talked. That is a mistake I hope to remedy. I suppose we have been at a point where our friendship is assumed.

A few fruitless months passed with Dan applying for everything even remotely related to his degree. Some were considerate and some rude in their response to his interview requests. None entertained hiring him and only a few granted an interview. More and more, Dan felt lost and depressed. Looking back after these many years, perhaps that period of time and those experiences were beneficial towards his growth. I say that as perhaps a way to reconcile thoughts and justify those bad times, but Dan was lost. He had no hope or foreseeable future. At least not a future he welcomed or one he would have predicted at his high school graduation. The time of focusing on a job search he didn't care about and a career he hated needed to end.

College debt was significant enough that payments had to start. Through a family connection, Dan got a job at a grocery store and received a decent wage. His impressive work ethic provided for frequent recognition which gave him a somewhat improved self-image. That enabled Dan to balance the disappointment he knew others felt at learning his degree juxtaposed with a current job as a grocery store clerk, but this job had no future that interested him. It never occurred to Dan that bagging groceries and chasing carts would be his career, but here he was facing that reality. The weeks and months since graduation slowly provided some level of clarity.

Something had to change, and he needed to be the one to initiate the change. Dan's parents agreed, but passively. He and another equally desperate friend occasionally considered options and commiserated together over a beer. For Dan, his thoughts were coalescing around two options as ways to jump-start his life. These two friends don't see it yet, but their direction in life might not coincide.

In time, things would change in ways Dan could not have predicted even two months earlier.

She passed the time…

Junior and senior years were very similar. She was conscientious and studied hard, but good grades were difficult to obtain, although she did pretty well.

Graduation, followed by a "real job" was her plan. Styling classes proceeded, but sometimes filled with drama as her instructor seemed to attract it, but the idea of graduating with a marketable skill and eventually becoming independent was almost intoxicating.

Others identified her by her group of friends as much or more than her own personality. Her group included both girls and boys. Her parents found comfort in her friendships, as they believed it provided safety in numbers. They usually met and travelled as a group. Sometimes they dated each other, but also broke-up without consequence. She was currently involved with one of the guys, but not in an all that serious way yet.

Daily life was honestly a bit of a bore. She wasn't challenged by school or work. Junior year turned into senior year and she developed a short-term plan. Grandma lived alone, but not for much longer.

In time, things would change in ways she wouldn't have predicted.

She welcomes your time,
Will you run or be a partner in crime?

Chapter 2

Didn't See It Coming

Thanks to his parents, Dan lived at home and had no expenses except for that college debt. His savings account exhibited small growth, but frequent 'beers with friends' just exacerbated Dan's depression, slowed his savings rate and gradually was becoming a defeating habit.

Drinking with friends provided short-term happiness though. This was one skill learned and developed in college. I now wish I had said or done something for him back then. What that would have been, I don't know. Quite a few months had passed since graduating. Dan and those friends developed a routine involving a few bars that they frequented. Although not 'regulars', they certainly had that routine.

She was beautiful and smart and yet she chose me. That's how Dan described it to me when they first met. The smart part is, I think, a reasonable assumption. She was a nurse at the children's hospital. I believe it's also reasonable or hopefully can be assumed she had empathy and understanding among her attributes.

How this night would change Dan could not have been predicted. The crowd grew at a predictable rate. Luckily, their routine provided for a timely arrival. It was a cool night with a clear sky. Stars brightly lit the night sky with the crescent moon appearing slightly larger than normal. Dan and his friends parked about a block away and walked the narrow alleys to the bar. This part of town is old, the bar and the others like it in this area were best described as neighborhood bars. None were trendy. Houses lined the narrow and occasional cobblestone streets.

On this night, they sat outside among the stars on barstools that gave them a height not provided by their DNA. Dan and his two friends were crowded around a tall, but small round table. Drinks and for some reason an excessive amount of napkins were placed before them. Their established pattern to be at this bar at this time was normally of no notice – that would change. The venue demanded patrons develop a comfort with a lack of privacy. The common area for patrons was small, partly because of the area reserved for an acoustic band squeezed into the far corner.

The outside area was encircled by ivy-covered latticework with no overhead cover except for the stars and moon. The bands typically played easy-listening music. Tonight was no exception-it was a couple of musicians playing acoustic guitars. The table of women to their side heard the men's conversation and the guys heard the women's. The inevitability of the moment is apparent only in retrospect.

The three gentlemen came to this bar and night out definitely not "dressed to impress". And since they were also not driven to physical activity or following trends, their look was mediocre – to put it gently, they were comfortable. Only notable among them was their joy. Thankfully, the band hadn't begun yet.

Their laughter enveloped half of this small establishment. Especially notable was Dan's infectious laugh. It spread between them and among the surrounding patrons. Strangers would accidentally let a chuckle escape while leaning in, but also left them wondering why they were laughing. This was a night like no other. The confluence of imagination, creativity, attitude and bravery validated and further cemented their friendship. The preponderance of napkins was the vehicle. The men's festiveness unexpectedly, but also inevitably spread to include the ladies at the next table. Three men at one table and three women at the next table. Hmmm....

So one might wonder what the source of frivolity among the three men at that tall table was. As previously mentioned, the number of napkins left by the waitress was excessive. The napkins and a pen willingly provided by said waitress provided the spark that made the night unique. Dan was the oldest among them (by only months) and the one to initiate the fun by creating a joke, writing it down and passing it around. All of which was very out of character, but these three were very comfortable around each other and supported each other. It was stupid humor, really, but they all laughed. It was honest levity – not an obligatory chuckle. Then the next did the same. Then so did the third of the men. This had never happened and before long nearly a dozen jokes were scattered and stacked on the small tabletop.

They couldn't contain themselves. Their laughter was so intense that it brought tears to their eyes and prevented them from enjoying their beer. People around them laughed without knowing why. The women at the next table eventually had to know. One of them tapped Dan on the shoulder to ask what was going on. Their jokes were shared, chairs moved so the two groups of three could become one and it suddenly seemed as if the six had been friends for years. They exchanged stories, drank plenty, and laughed often. This night was most unusual. And as time would tell, not predictive of events to come.

The night at the bar eventually came to an end. Dan was dreading this. The girls decided they wanted to or needed to go home, but in a strange way, I think Dan was glad they decided to go, for it meant they would separate on a high note. Their memory of the night would be positive he believed. It was certainly a welcome change to anticipate a nice memory. Dan needed a night out like this. And yet he was nervous for what it meant. He would have to quickly find a way to ask for her phone number. It would be at this point he would know how she felt about him singularly.

She quickly found a pen and paper (napkin) and enthusiastically penned her name and number and asked Dan to call. He walked on air the rest of the night. Dan was not good at meeting girls and making conversation. His two friends jokingly reminded him of this after the girls left, recounting his stumbling efforts. As the days and weeks went by, she and Dan phoned each other occasionally. Eventually though, their date was scheduled.

Dan was taking her to a concert to see one of her favorite small bands. They would join her friends. A group of about five couples. This sounded like a perfect first date. Dan felt relieved of pressure. Besides, he also liked the band and dinner beforehand was at her apartment. It was indeed perfect! Looking back now, it's almost unbelievable. There was so much promise. Somehow, depression always ruins everything, but this time Dan had hope again.

Their personalities blended well. I even went shopping with Dan for new clothes that he'd wear to impress her. That effort was exceedingly rare. He was happy. I was too.

Walking up to her apartment, Dan was nervous, excited and confident. He had red roses delivered that day. Ascending the steps to her front door meant passing by the garbage cans. That's only pertinent because of the bright red roses peeking out from under the lid of one can. Dan's emotions became suddenly confused. He had a sudden sinking feeling – it was that lost, looking into the distance type of experience. Her opening the door and welcoming him in snapped Dan back to the present. He managed obligatory comments in response to the tour around the room to meet her group of friends. His facade needed to last until his eyes could verify that his thoughtful gift was indeed the refuse bouquet.

Dan offered the standard responses and expected smiles as he interacted with her friends. She and Dan had minimal interaction before or during dinner. Drinks and the meal became barely a memory except that those roses were indeed what he bought. Dan couldn't conceive of a way to ask why or how.

The concert was festival seating, so there were no assigned seats. They were all forced to sit in one long row with Dan on one end and her as his date beside him. Drinks were in hand and ready. The band took the stage.

Perhaps symbolically or ironically, a string on the lead guitarist's guitar snapped during the first song. The night didn't start as the lead guitarist planned either.

He got a different guitar and the band started the song and concert over again. Dan didn't have that option but was optimistic. The entire concert seemed a wonderful success for his date and her friends, though the cliquish companions left him disregarded. They didn't need him and most importantly, she no longer needed Dan on this date.

I see it all as clearly now as when Dan first described it to me. All her compadres were couples. Dan filled a one-night need. The music helped him get through the night. After the concert, the group decided to walk to the bar across the street and end the night with a drink. She mentioned to Dan that he didn't have to go if he'd rather not. She'd just get a ride from a friend. He was quite aware she didn't need him anymore.

He, of course, parted ways with her eagerly. Dan never expressed anger, but often his silence was indicative of it. The two never spoke again. Surely he didn't wish to, but probably she didn't either. She was beautiful and smart (and deceitful and manipulative) and yet she chose him.

Everything in Dan’s world seemed to be crashing down again.

Friends from college were becoming more and more

distant and his future darker. Would Dan become an embarrassment to even himself? A belief of not being loved was not as bad as being used. The days were definitely darker and the future bleaker for him now.

She wanted change…

The girl in that town to the north finished high school, achieving her stylist certification and moved in with her grandmother. Her plan was coming into focus. Soon she hoped for a position in a salon, but she didn't quit the fast-food job just yet.

Living with her grandmother wasn't all that bad. Rent was minimal. Her independence was achieved, relatively speaking, with her comings and goings never questioned anymore. Grandma was a very easy-going sort.

However, the income wasn't very good at the salon position she eventually accepted and it was led by a relatively sexist person. She wasn't permitted to use her skills cutting hair just yet. Instead, she was relegated to shampooing heads of hair all day, every day. There was no future ahead if things didn't change, but because of a five-year contract, she couldn't leave for another salon.

She currently had a skill that didn't bring her happiness and little options to pursue. Work life was seven days a week since she was working two jobs – the salon and the fast-food restaurant. It wouldn't be accurate to say she was depressed though. The guy she was dating wasn't perhaps "the one" but was nice and with a nice family. She did enjoy his company and maybe a more serious relationship was possible. Plus, she still had that group of friends to enjoy the weekend nights with.

This time of her life was spent treading water. Her pay was considered small even in those days and there was no way could she afford her own apartment. In fact, the next five years looked bleak. She decided to save as much money as possible to go on a nice vacation in the coming years. That much was under her control.

Every day was just living the status-quo.

Take it easy, let the blues come through.
Feeling low is normal sometimes too.

Chapter 3

Out With The Guys

A year has passed since Dan graduated college and was gifted his diploma by that chemistry professor. The idea of using the degree has long-since passed. Cold days have returned again and most of the friends from college are seemingly distant history.

More and more Dan reminisces about summers between the four years of high school spent backpacking in Daniel Boone National Forest. Back then he had friends, people generally thought well of him and the future was full of attainable goals and exciting options. Heading again to that forest with no intention of returning is an unreasonable but considered option. Surely, no one would miss him.

I was worried but felt helpless.

During one of those days reflecting on his depression, circumstances changed due to an unexpected phone call. Dan's college buddies were planning a camping trip now that spring was near and wanted him to join in. Dan realized this was only to be a brief relief from his current life but was eager to jump at anything. Hope being given and taken is now an experience he knows well.

That bond between college and fraternity buddies will be forever unbreakable. They, on some level, always knew they would someday go their separate ways, but college experiences allowed them to ignore this until well after graduation. The camping trip being proposed is one they'll dutifully and excitedly attend.

At this point, about a week and a half remained until the day they would leave. Jason would be leaving to start his new job in a new city soon after the trip. The camping trip was their way to wish him well and send him off. The date for his departure and starting his new life couldn't be changed.

They'd say goodbye with a week-long camping party in the woods, but shortly before the planned departure for camping an unusually late-season snowstorm enveloped the area. It wasn't a blizzard, but roads were relatively hazardous. Apparently, their expectations of spring weather weren't realistic. They met to talk it over, but the truth is none thought it was a good idea to challenge the storm except for Jason and apparently his mom. They met down in Jason's basement to talk and eventually decide.

Jason's mom had a lot to say and a it had a big impact even though she wasn't invited and didn't attend the meeting. Her plan proved flawless. As a few of them sporadically left the basement discussion to use the first floor bathroom, she'd stop them before returning to the basement. She let each person know how much Jason needed this trip. She told them how important their friendship was to him. She stopped just short of begging each person to re-consider, brave the elements and go on the trip. And besides all of that, she held the title of Mom.

I guess each person thought they were the only one she stopped. Not everyone went upstairs, but each who did changed their mind. All aspects of the trip were given its due consideration. I mention that because it's rare for this group of fraternity drinkers (I mean men) and former college partiers (I mean students) to show responsibility, but it almost happened. In the end though, with very little planning, they chose to disregard levelheaded decision-making and go camping. I wonder if Jason's mom ever told him.

The decision to proceed with the trip proved to be a good one. Weather cleared, meaning there was no active precipitation. The temperature was cool to cold, but they could prepare for that and the nightly fire with drinks helped as a distraction. Several days of day-hiking, drinking, and laughing were interrupted only by a search for an urgent care clinic. The entourage just wanted to get it over with. Dan, as the limping leader probably felt the same. He put aside the thought of, *Why me?* In relaying this story to me in later years, Dan said on that day he was deeply embarrassed at what seemed liked proof he couldn't do anything right anymore.

Dan was limping badly due to the pain he denied to his friends. His pants were badly stained around the lower part of the left trouser leg. That same stain appears to have soaked his sock and shoe. The group of friends dutifully followed. None seem concerned, as their destination was in sight. The clinic was soon to have guests; the entourage relieved as their limping friend was on the verge of getting help – and doing it without saying a word. His eyes were expressive.

Doc and Eve just caught sight of the approaching group. Doctor Evans and Nurse Stevens (Eve, as she's apparently informally and usually known) have just finished an unusually busy day. Sunset is coming soon as they depart their small, homey clinic on the edge of the woods. Daylight is diminishing as she finishes by turning off the lights as she leaves. Doc dutifully, and habitually holds the door as she approaches. They look at each other and sigh. Neither uttered any words – none were needed. Both clinicians went back into the clinic, turned on the lights and proceeded to provide care. Eve had a lot of questions for Dan.

Be outgoing with peers and with others be shy;
and be careful with people who know your thoughts
when you sigh.

Dan mentioned those words to me as something he always gave consideration when meeting new people. Sometimes he got a little poetic in expressing thoughts or opinions. In this case, Eve was that new person. Dan felt it important to only provide the information she actually needed. Eve's many questions, coupled with the feeling she was developing some notion of Dan's mental condition, was becoming uncomfortable I think.

Dan, his group of fraternity buddies, and Eve crowded into one room in the clinic. There was never consideration that Dan would be left to deal with this alone. Raising the trouser leg revealed the source of what stained his clothing. It was the result of a relatively deep mishap caused by a misguided hatchet. While chopping wood for a fire back at camp, the angle of the device was allowed to become too shallow, resulting in a glancing action that caused it to veer off course and into Dan's shin.

Eve was unfazed as she gathered the necessary supplies. Doc would also join this crowded room to confirm her diagnosis. It looked a mess, but easily repaired. Neither Eve nor Doc seemingly thought it to be much at all. And so neither did the guys. It became only a minor hiccup – a short story which will receive the obligatory embellishment in later tellings.

They would shortly be back at camp where they sat, drank, talked, and laughed. Beer provided the pain control Dan required. This group had formed a strong bond over the years as fraternity brothers. I must stress they were *brothers*. This past week further strengthened an uncommon bond. No one need tell them that. Hugs were neither uncommon nor uncomfortable. However, unfortunately and unpredictably, this group would not gather again for thirty years – so sad.

Obviously, the trip had to come to an end. It was a wonderfully fun and needed break. I think drinking around a fire during those nights away in the forest with that group of special friends helped Dan. He must have, I believe, opened up to them. At least as much as Dan could open up to anyone. The memory of that trip kept him afloat and he would often refer to it. Nothing in his life had changed, but Dan somehow seemed less forsaken.

The girl to the north maintained…

Four hours north she maintained the work ethic people envied and respected. Her goal to save funds for a vacation was happening. She was as persistent as anyone could be. Once she defines a goal, it would be stupid to bet against her. She needs more time, but she now felt comfortable thinking of possible destinations. Her only requirement was a beach.

The salon gradually loosened their requirements and gave her limited hair styling time. That meant slightly more pay. She maintained her hours at the fast food joint. Grandma was encouraging and held her to no real hard and fast rules. Life was comfortable though not really leading anywhere. She did not indulge in long-term hopes or goals, but her everyday life with her boyfriend wasn't so bad. By this time, they had been going-together for a couple years, but still weren't at a point of contemplating anything long-term.

If life was going to change (and it will), it would happen to her instead of because of her. At this point in time, six years of age and only four hours of distance separated her from Danny, but neither knew the other existed and no one would have guessed their futures.

Dan considers a future…

During the transition of spring to summer Dan had privately decided on a plan that he had discussed – and I thought seemingly forgotten – from months previous. The Peace Corps or the military interested him as options to provide his way out. Grocery store clerk was just never how he saw his future. Dan related to me that at the risk of being perceived as conceited, he always thought he was going to achieve something.

The Peace Corps provided the opportunity to experience different cultures and help those in need. Dan would do things no one he currently knew could ever relate to. He might be deployed anywhere around the world. The down-side as he saw it is that in ten years, he might not be any further along in achieving anything. What would Dan do when he tires of those extended trips? Would there be any place to call home? Would there be anyone he'd call a friend for longer than the current project work?

Dan was starting to sense an obligation to serve, but it might have to be in the military. His father and other family members served. They all turned out okay. Surely there must be programs that provide training and jobs that would be applicable to the civilian world after serving a term, but all these plans would get their due consideration in time. Something just came up that's a lot more interesting.

We'll take for granted the people we'd miss the most.
To friends long lost, we offer a toast.

Chapter 4

Decision Time

The time since and his experience following college had been far more difficult than Dan expected. Honestly, I expressed this to his parents not long ago. It is my conclusion that the past drives perception and perception defines success. However, since all this is long-since past, perhaps my opinions ought to be left to myself.

The only expectation placed upon Dan was to attend and graduate college. No performance-based opinions were discussed or associated with college performance or post-college actions. His parents just wanted Dan to be comfortable financially. No consideration was given to enjoying some career, but it was his assumption he'd meet any expectation – even those he assumed of others, but his achievements would come only after he surrounded himself with a very different environment than he ever gave fair consideration.

The Peace Corps, Dan decided, would be for him a short-term experience, but while certainly rewarding and definitely helpful for personal growth, as mentioned previously, the chance existed he'd be right back to his current situation afterwards. Dan felt he needed something more substantial to "snap him out" of his depression or whatever experts would diagnose as his current condition. Investigation into the military would take time.

His methodology would be to talk to a recruiter from each branch, minus one, and hopefully discover a path for him to pursue. While making appointments with recruiters, he'd continue as grocery store clerk and increase his exercise regimen to be in his best physical condition for boot camp whenever/wherever that would be. Dan felt good about his plan.

The Marine Corps was immediately eliminated and the only recruiter Dan did not visit. He did not envision himself ever feeling the intensity towards anything like the Marines did toward the Corps. Additionally, anything infantry-related was of no interest. So visits were arranged with the Army, Coast Guard, Air Force and Navy during days off from the grocery store.

Meanwhile, the grocery store chain was implementing a program for staff to voluntarily work occasional shifts at other stores to help address staff shortages. His strong work ethic made him a target for management and he honestly didn't mind the attention and positive words. Besides, he got this job due to a family connection. That made it impossible to refuse favors, even from management, but it's also true he felt this might help distract him from his current feelings about himself so he accepted a few 'voluntary' assignments. His assignments were all at the same store and the work was the same as his current position so adjustment would be easy. The change of pace was welcome. Plus, seeing her face among the new faces was enjoyable.

The store Dan was helping was across town in a more affluent area. The average income of the customers didn't matter to him. After all, Dan didn't have much interaction with customers – he stocked shelves, unloaded trucks and organized the back room storage area.

Dan and his brother went together to the Army recruiter since his brother was an Army veteran. The thought was that his brother would help decipher the recruiter's sales pitch.

"My Dad, several uncles, and my brother here all served in the Army. I'm considering it as well and I'm wondering what my options are. I plan to talk to someone from each branch, but wanted to start with the Army," said Dan to the recruiter.

That's how he started the conversation based on his description of that day. The recruiter, after a few questions, learned Dan was a college graduate and immediately recommended pursuing the Army as an officer.

But Dan absolutely always avoided the limelight. He felt being an officer in the military put the spotlight on him. In fact, he wasn't comfortable with even the smallest of recognition in front of a group. I know of circumstances when Dan requested an award be presented privately in an office somewhere with the door closed.

It's complicated.

I never believed there was a lack of self-confidence, but certainly one who truly knew him well might describe a diminished self-worth. Dan often said he likes to be in the background. "Everyone can just do their own thing," he once said. "I'll be fine. Just pretend I'm not here."

He never felt the need or importance to be the leader in order to lead. So becoming an Army officer did not suit his interests. He'll give honest effort to tasks that are required, but recognition for his efforts are not required.

The recruiter pushed the officer program to exhaustion. When Dan says he's not interested, it's best to just move on. Not doing so will just piss him off and he'll shut down. His sentences will get short and words more pointed.

The recruiter did not respond well to his efforts being rebuffed. The distance between them grew as their sentences shortened. The recruiter must have known (as Dan definitely did) that Dan would never be a soldier. He left the meeting sure that one direction his path would never lead was the Army. He was miffed if not angry as well as disappointed that this initial recruiter visit went so badly. Dan put up a blockade that future recruiters will have to deal with if they want his services.

The staff of the grocery store across town was helpful and quite nice. They seemed to appreciate the help but did not gush. That suited him well. He didn't want anyone to make a big deal of the help given them. No matter what anyone's opinion was, he just wanted to enjoy the change. Dan kept busy doing whatever was asked of him. He interacted as needed but sought only to meet one particular clerk.

She also stocked shelves, but they were often working in different aisles. In retrieving more items from the back room, he'd find a way to pass her, saying hi. Initial conversations were short, not substantive and casual. She was pretty and Dan thought easy-going. Somehow she came across as amicable and yet evasive. Her height was slightly less than his with unstyled shoulder length straight brown hair parted off-center. Facial features were soft with eyes also brown. She seemed to enjoy and hopefully welcome his awkward attempts at conversation. She was approachable, but not outgoing. The days Dan worked there were irregular and unpredictable – no set schedule. Their casual conversations were eventually more involved, but never of significant depth. It's of course very difficult to have a meaningful conversation while stocking grocery store shelves. He was stupidly shy with her, never asking for a date or her phone number but did get her name, Ellen.

Unpredictably, an unusual surge of bravery arose on a day off.

The visits to the Coast Guard and Air Force recruiters were similar. Neither of the recruiters were pushy. It's speculation to be sure, but I believe Dan wasn't joining the military to just satisfy an obligation or jump-start his life (as he said). Sure, Dan had to do or change something. In just months he'd reach the two-year anniversary of graduating college. He needed a change to something he did well – something to make him feel accomplished. Life, as it was, wasn't enough if it was currently at its apex. I believe Dan was searching for self-respect and he knew he wouldn't achieve that at the grocery store.

The Air Force had the opportunities to offer. He'd be able to see places he'd probably not go as a civilian. Unfortunately, Dan's eye vision, or lack thereof, wouldn't allow him to pilot any craft. His good friend from high school was a mechanic and liked it. Dan wasn't interested in that and there didn't seem to be much available from the Air Force that he couldn't do as a civilian. That was one part of his criteria for the military. Additionally, the recruiter didn't even try to push any unique locations for Air Force bases.

His friend was stationed in the middle of nowhere in the Dakotas. He didn't want to chance that happening to him. Sure, Dan wanted the military to give him skills transferrable to civilian careers, but he also wanted experiences that civilians can't experience. The recruiter didn't pursue him. He thought that to be unusual. So the Air Force wasn't eliminated but was also not preferred. Dan commented to me once that he expected the recruiter to do more to get him to join. Perhaps his expectations weren't appropriate.

The Coast Guard was interesting in that life attached to a sea-going craft was different enough. At that time though he couldn't be guaranteed a duty station location. And the available jobs as described by the recruiter weren't impressive. The future as Dan saw it would be unpredictable at best. He couldn't risk leaving the Coast Guard someday only to return and relive his current condition. This recruiter, as with the Air Force, didn't seem to want to take the time with him.

This decision was becoming more difficult. There was no clear choice. Dan was starting to feel self-doubt again about himself and certainly his plan to move life forward. What if neither the Peace Corps nor military could give him a future? Dan hadn't considered his plan completely failing.

The options for the change he envisioned were not

producing a clear choice or path. He thought about the advantages and disadvantages of each constantly.

Dan had a day off from work and relatives were coming over to celebrate something. They will ask about Dan's future plans. He's at home but doesn't want to be there. He could leave to go somewhere but can't think of anywhere to go. Dan was just tired of this life and he needed change just for the sake of change.

If December comes and nothing has changed, he might take action everyone would reflect on, saying they didn't see it coming, that they are surprised he did it and would have helped had they known.

Hope is fading fast, but Dan will talk to the Navy recruiter anyway. It was part of his plan to do so and he felt an obligation to himself to follow through. Dan had no interest in returning to school for some type of technical training, but that might become his best option. The thought of that made him seek the comforts of a drink.

"This whole thing sucks!" said Dan to me while we shared one of those drinks, but the Navy recruiter was quite upbeat at hearing what Dan had to offer and what Dan wanted out of the Navy. Certainly, the Navy was more friendly than everyone else he has visited. They also provided unique opportunities and jobs. Not only had no one in the family been a sailor, but serving aboard combat ships was definitely unique, and if he was to serve in the Navy, submarine duty was Dan's preference. Could he choose to serve on a submarine? And would he gain skills and training transferrable to civilian life after his enlistment? Yes, and yes. The catch was that he'd have to agree to a six-year term instead of four, but he'd be on a submarine! No one Dan had ever met could say that. All his boxes were checked. Dan left without verbalizing a decision but took the recruiter's card. Anyone could see Dan would be calling. Maybe it was a blessing that his paper technology degree didn't turn-out.

Dan was of course friendly to his aunts and uncles as they arrived. No cousins or anyone of his age were there or coming. It didn't happen often, but Dan would have rather been at work. The thought was impulsive, but Dan decided to drive to that grocery store across town. He was nervous and distracted during the drive.

If Ellen was working, she'd be getting off soon after he arrived.

He parked and walked the aisles searching for her. If he looked out of place, he neither cared nor noticed. Towards the end of one aisle Dan spied her bent down stocking a lower shelf. His pace to surprise her was slow, but deliberate, as he was deep in thought. Dan didn't know what he would say when he reached her. She looked up, sensing him when he was near. Both smiled as they greeted each other – he thought her eyes showed surprise and happiness. Dan explained his hasty decision to drive there. Both were aware of the lengthy drive required by him to 'stop by' and they both understood the only reason was to see her. Dan's confused emotions were either infatuation or something more, but before he would try to decide, he wanted to see if she thought he was worth it. In little more than a month he'd be leaving for the Navy. She welcomed him being there without saying a word. She'd be off work in a few minutes.

They decided to go for a walk around the nearby outdoor mall.

The Navy recruiter had called with another option. Dan could be a “nuc” (shortened nickname for nuclear technician) if he scored well enough on a test. A “nuc” worked in the engine room on subs operating the nuclear power plant and had excellent employment opportunities upon leaving the Navy or fantastic bonuses for re-enlistment. Dan scored quite well on the test so he had several excellent options if he enlisted, including that as a nuc.

Dan did indeed enlist but chose the path of ballistic missile fire control technician rather than nuc. He'd be given electricity and electronics training and work in the missile control center of a ballistic missile submarine.

Excitement doesn't begin to describe his feelings. In time the grocery store and the useless college degree that haunted Dan would be behind him. He was now determined to make something of himself. This choice was entirely his. Dan felt that for the first time in his life.

Attending college and selecting a paper technology degree wasn't his decision as much as it was what was expected. If anyone doubted this decision to join the Navy, Dan was determined to prove them wrong. Working at a grocery store wasn't so bad now that he felt he had a future to pursue, except that the staffing problem was resolved so he wasn't needed across town anymore. Therefore, he probably would never cross Ellen's path again. That bothered Dan.

The temperature outside at that outdoor mall with Ellen was comfortable with a light jacket, as he relayed the story to me. The breeze was slight and the sky was mostly cloudy with daylight beginning to fade. The outside mall lights marking the walkways were lit due to the approaching darkness of evening and the cloud cover. It was crowded since it was a weekend evening. People rushed past each other without taking notice.

From a distance, the two of them looked out of place among the crowd. They strolled slowly without direction and talked. Dan desperately wished to hold her hand, but lacked the bravery and didn't want to impose. She declined food and drink. Conversations were easier to him with an adult beverage – he struggled on this night. She did not indulge him when asked about her life. Perhaps shy, modest or uncomfortable with her story, but he thought her to be suddenly nervous so he never pressed the topic. She'd change the topic to ask about the military service coming soon.

Dan thought he sometimes had no luck if not for bad luck. The military life he was about to start would change his life.

That's what he intended. Life had honestly been mostly downhill since college. Naval service was a pursuit of change. One that would change his life and lift him from his funk – it would provide milestones. That was his expectation, but also true even in hindsight.

However, maybe Ellen could have also been that change he needed so badly. He certainly believed it possible, but he had already signed the papers for the Navy so there was no turning back. Their steps eventually led them back to where they started. An hour of conversation and casually strolling passed quickly. At his request, she provided an address so he could write, but for now she said she needed to get home. They were expecting her. She wasn't to be late.

Certainly he would return from Naval training at some point and asked if he could then see her. She happily agreed to that plan. Ellen and Dan could then meet and catch up. With that they exchanged a light, but friendly hug and said goodbye. Each went to their respective cars and drove off. There was something different about her.

Dan later admitted to his eyes welling up as he drove away due to thoughts that this might have been the last time they ever speak. If only he knew. The long drive home allowed time for him to straighten up and ensure no evidence of his emotions were present upon arrival home. Upon arriving home, Dan was asked where he was for so long and he said he ran into a friend. That satisfied the inquisitor. He was already missing her and so started thinking about writing a letter to her from boot camp. He assumed he'd be allowed that privilege.

Meanwhile, in that town to the north...

The girl treading water as a stylist is close to her limit. She is now trusted to perform limited styling, but this job is taking way too long to get to normal. The money she needs can only be achieved by styling hair a lot more than she's currently doing and working two jobs has been quite difficult for this length of time. And styling hair is quickly becoming just a job, rather than an evolving career. Her personal relationship has stalled. Living with Grandma hasn't been a hindrance, but she wonders more and more what true independence and freedom feels like.

She, like Dan, feels she has no future. She wonders if she did something wrong, but what would she have changed? She doesn't have regret. At least not yet.

Working hard and being nice hasn't gotten her anywhere. She is considering leaving both of her current jobs for one job at a local drug store. Her friends encourage her to sit tight – things will change, they said. That's okay for now. She is patient and doesn't like change so sitting tight is tolerable for the near future.

No one could have foreseen her future, but her friends were right, things will change.

Thinking a friendship is yours alone,
Makes its loss your burden to own.

Chapter 5

And So It Begins

My new journey is beginning soon. I did not expect these last two years to unfold as they did. In trusting confidants during high school, I guess I believed I was more capable. I thought I was smart enough to not just do well in college, but to do well in the most difficult major the university had to offer. After all, why would the college counselor recommend such a demanding course load if I wasn't right for it? But I guess I was either not smart enough or not capable of applying myself. It doesn't matter once the grades are in and graduation is complete. My low grade point average coupled with my complete disinterest in paper technology (my degree) sealed my future. I had no reasons to give potential employers to hire me or to take a chance on me. Perhaps someday I'll get a new degree in a field of my interest, but I have no interest or the money to pursue college or alternate training now or the foreseeable future.

Two years have passed since graduation. Two years of lost effort and time. I had no direction or focus, but I was smart enough to know I had failed, my life was uninspiring and every aspect of my life needed change. I can't think of anything in the last two years I want to preserve in memory. Somewhere deep inside I felt an obligation to serve, but how or who was absent.

I think that's how I'd describe my thought process.

Truthfully, that didn't drive my decision to join the military, but that need existed then and still does. There is always someone out there I need to help or serve. I decided to depend on the military to help me get out of this hole I dug for myself.

The weekend prior to my scheduled time to take the oath and start my Naval career my friends from college, some of which I thought were long-lost, surprised me with a going-away party. I met them at the apartment complex where one of them lived. We walked to what I learned was a party room, upon opening the door everyone yelled "Surprise!!!"

I am not a fan of being the center of a surprise party nor comfortable being the center of attention. I regret this, but when they yelled, I turned around, let the door close and started to leave. My friends behind me laughed and we re-entered as a group.

I had a great time that night reconnecting and laughing together. I realized then how much I missed all of them and how much I'd miss them in the future, but I also confirmed my lack of success. Most were working somewhere they at least tolerated and expected a bright future. My bright future is delayed, but it seems I still have people that are friends.

Maybe I never really have been all alone, but I also knew after that night we'd go back to the lives that caused us to be separated. That's just the way it goes. In the future, perhaps I'll remember to commit myself to maintain a connection with friends instead of assuming it exists.

In college I was in excellent physical condition due to my efforts as part of the men's gymnastics club. I maintained that (to a much lesser degree), but perhaps increased my cardiovascular conditioning during the last two months. That's probably all I did well. Everything I heard about boot camp in the military stressed the importance of physical conditioning. Now the time is near for this milestone – me joining the Navy.

I'll sleep tonight then start my six-year commitment tomorrow. Sometime during boot camp I hope to write Ellen. Somehow she means a lot to me. I miss her and yet we've hardly spent any time together. I don't think anyone would say we even had a relationship. We haven't even had a single date. These emotions are confusing.

I and my parents woke before sunrise thanks to an alarm clock ensuring I wouldn't be late. They drove me downtown, I took the oath with half a dozen others, and said goodbye to my Mom and Dad and my miserable life since college graduation.

I met the group I was travelling with to Chicago and then on to Great Lakes, Illinois – the site for Naval boot camp. It's only a short drive north of Chicago, close to Wisconsin, but I immediately forgot the names and faces of those in my group almost as soon as I met them. My focus on boot camp was intense. I planned to turn boot camp upside down. I'll be the best damn recruit they ever saw. I couldn't wait to start. I was almost shaking with anticipation.

I didn't really know what to expect, but they also didn't know what to expect in me. I was ready to go so badly it seemed as if I was pissed-off at them. I would not allow them to break me. I thought, they have no idea what's coming their way. It was very early in January and very cold. A constant frigid breeze kept us alert, although no snow was predicted.

I and my fellow new recruits all rode in a seemingly unheated bus from downtown Cincinnati to the airport and caught a plane to a Chicago airport. A sailor was to greet us in Chicago and direct us from there. At the Chicago airport we boarded another bus for the extended drive to get us to our final destination. Up until arriving in Great Lakes, Illinois (boot camp) everyone was quite nice. No one had yelled at me yet. Of course, that was about to change. That was my expectation.

Boot Camp, day 1

Upon arriving in Great Lakes we were ordered out of the bus and commanded to stand in line at attention. Of course we did it wrong – we didn't know how to yet. We all knew we'd screw it up. That's when the yelling started. Nothing we did was right. I believe they had practiced and planned on this. We were expected to be overwhelmed, confused, and unsure, like probably everyone in the past. I was almost looking forward to this moment. It was all an act, I believed, but at least it made me forget about the cold (which was admittedly significant). Boot camp was meant to break us down and rebuild us in the Navy mold. This wasn't shock and awe to me. It was all short-term. Yell at me and make me do pushups-who cares. In eight weeks it'll be over. I would never acknowledge them as anything more than they were – teachers with a style none of us were accustomed to, teaching a subject we have never been exposed to and performing an act well-rehearsed.

We had to relinquish any belongings we travelled with, told to stand in line (in silence) and wait for our turn to have our hair cut. As each man had his hair cut almost too short to be felt, he had to wait for everyone else. Everything was done as a unit or unified group. We were now Company 008 – no longer individuals. We looked like different people now with no hair. Once everyone's hair was cut, we were told to assemble, at attention. Eventually we'd get good at it, but not yet.

With heads shaved, we were told to march single-file into the building next door to get clothing and supplies. We were first given a sea bag. Moving along the line we were supplied with everything we'd need. The clothes worn to Great Lakes would eventually be taken and packed away, not to be seen until the end of boot camp. Or in the case of my hoodie, never to be see again. We received Navy-issued clothing and supplies, everything stuffed into our sea bag. What a mess! I thought, screw it, just go with the flow. This is all by their design. It's not like I had a choice anyway. I had no idea how any of the clothing would fit. Anyone not knowing their size was verbally ridiculed. Thankfully, I did.

After filling our sea bag, we assembled as a group again before moving to the next required stop. The sea bag was heavy. It was comical as we all struggled to carry it and march. I didn't risk a smile, but I'm guessing the men directing us got a kick out of it. Each of us leaned as we walked, no one in unison and inadvertently bumping into each other as fog exited with our breath as we moved along.

Our next stop was to an area similar to a high school cafeteria. The large room was open- no stanchions. It was filled with long folding tables and chairs. We were to march down one side, drop our sea bags and then fill in the chairs from front to back. They yelled at us the entire way. We marched wrong and sat wrong apparently. Some relaxed after sitting – that was wrong. I guess we should have known to sit at attention. Who knew that was even a thing?

It was at this point I considered that everything we did for the next two months would be wrong. Even if only one person in the eighty person company did something wrong, we all would pay for it. What they demanded from us was to perform orders exactly as told, nothing more or less. At those tables we completed a lot of paperwork. When told to write our names – do only that! Do not fill in the birthdate until told to do so. My thought at this point is that I should always strive to just blend in. Just blindly follow orders. I did nothing to draw attention. I didn't want anyone to know my name. The time was now after midnight. It had been a long day and it didn't appear as though it would end anytime soon. None of us could relax. The onslaught was constant, by design. They had eight weeks to turn us into a cohesive group of sailors.

We were then given black markers and stencils to be used on our just issued Navy uniforms. We retrieved our sea bags and sat back down. Shirts, pants, belts, hats, coats (almost everything) needed our last names stenciled in black or white, as directed. Additional miscellaneous items included the mesh bags used for toiletries and the sea bag itself. Surprisingly, socks and underwear were excluded. There was specific direction as to the location on each garment for the stencil. This task seemingly took forever.

It wasn't difficult if we did exactly as told, but we were tired. Some placed their stencil incorrectly and were again ridiculed. We eventually learned we would be verbally harassed throughout boot camp for our stenciling mistakes. Surely, some mistakes were simply due to the harassment and by now, being very tired. We were exhausted, but this wasn't our last stop. Sea bags were repacked, hoisted on our right shoulders, and Company 008 reassembled.

The next step for us was our barracks. This was another huge room, but this time filled with bunks stacked two-high. Separating each set of bunks were lockers to which we were assigned based on our bunk assignment. We barely enough room to store our new belongings – no locks needed. There was nothing to steal, minus toiletries. At this point, literally almost everything had our names on it. I don't remember how we were assigned to a bunk/locker combination, but I found myself on a bottom bunk at the front of the barracks. Not being tall, I was happy with the bottom, but wished for a bunk in the rear – especially as time went on. Being at the front of the room, my sleep would be disturbed by anyone entering the barracks at night or any fellow recruit using the bathroom through the night. I believe I grew accustomed to the traffic and also seemingly required less sleep than as a civilian. So the bunk location became less and less important.

Suddenly, two gentlemen in uniform entered the barracks – our company commanders (COs). We were commanded to attention. Everything they said was at an elevated decibel (they were yelling at us too). They introduced themselves, but also informed us that we were to only call them "Sir". Both were petty officers (enlisted personnel) meaning no one, including us, would or should call them "Sir" outside of boot camp, but different rules apply now.

One of them was a taller black man, around six foot three inches or so. He was somewhere in the middle between slender and husky, clean shaven with hair cut very close to his scalp. His uniform was well-pressed and shoes shined. The other gentleman was white, about five foot seven inches tall with light brown hair. He was also clean shaven and a burr haircut with the sides of his head cut extremely short. His build was husky, also with a pressed uniform and shined shoes. They spent this time yelling at us, something about being our worst nightmare if we messed up. I didn't feel tired with all the stimulus being provided, but it had been a long day. It was well after midnight. At some point lights went out, we were told to "... get some shut-eye" and our COs left.

It was now 2am, maybe even a little after. The first day of boot camp was over. It was a long day, but it passed quickly. It seemed so hectic I had no idea what to expect next. Everyone I asked prior told me there would be a lot of running and calisthenics.

I had just fallen asleep when it seemed all hell broke loose!

Boot Camp, day 2 and beyond…

Our two COs snuck in, or perhaps I was too tired to be aware. Each took a lidded metal trash cans and threw them down the middle of the barracks and began yelling as the garbage cans bounded off the floor and railings at the foot of the bunks. That noise startled everyone, causing recruits to jump from bunks – with some forgetting they were in the top bunk.

The time was 4am. Imagine eighty or so men in white underwear and t-shirts jumping from bunks, scrambling to make their bunk, get dressed, and run to use the restroom and shave. And I do mean we were running to and from the restroom. The barrage of orders directed at us was intended and did have us overwhelmed, confused, and scrambling.

"This is what you are to wear today," shouted one of our company commanders.

He then read a statement listing the "Uniform of the Day". He read it this day and then various recruits took turns to read it each day of boot camp after that. The uniform of the day listed the entire uniform we should wear: the pants, shirts, shoes, hat, coat, and gloves as required. Since it was January and we were just north of Chicago along the western shore of Lake Michigan, we always wore a peacoat, gloves, and knit hat when outside. It was mayhem in the barracks.

Most of the morning of day two was in the barracks receiving instruction as to how to make our bunks and the need for speed getting ready in the morning. This was not teaching from an empathetic, considerate, and patient teacher. Rather, we received our lessons from an irate individual yelling at us as a group and occasionally individually from a face-to-face distance of only inches.

I assumed and believed everyone knew this was all an act, but their methodology did allow me to quickly learn how to make a bunk with military corners quickly. Doing so meant avoiding the attention of a CO. They also told us to fold our blankets and place them under our mattress for a night. That would create creases, they said, so that "The blankets will almost fold themselves". I followed that and other suggestions without question. They were right. My blanket took only seconds to fold each morning. Surprisingly, dozens of recruits didn't comply and it showed in the bunk during inspections. It wasn't unusual to have their blankets and sheets ripped from the bunk and told to start over during the occasional morning inspection. I thankfully never received such treatment.

I don't remember if it was day two or subsequent to that when we received our immunizations, but the event was memorable. Eighty men facing and leaning against the outer wall encircling a large room. Very little furniture sat in the middle of the room, but an echo of loud, verbal orders overfilled the space. Individuals came by to vaccinate in either shoulders or butts.

The memory of a roomful of men with their pants down is an unwanted memory. "Don't move," we were cautioned.

I saw the result when some anxious person did. The vaccine site turned into a small laceration. At least for this short period the yelling was less intense.

We received some sort of instruction every day of boot camp. Most often it required marching through frigid weather from the barracks to a classroom type of environment. The cold kept us inside for everything. In fact, we spent more time in classrooms than running or doing calisthenics. That was completely unexpected. We marched as a group of eighty men everywhere – to the gym, dentist, and classroom – everywhere. A group of eighty marching down a sidewalk two or three abreast resulted in a rather long line.

Each time we approached a street to cross, the crossing guards were called to the front to stop traffic. The crossing guards marched at the rear of the group and were required to sprint to the front when called upon. I was one of the crossing guards.

We ran more than anyone and were also the last to enter the cafeteria (or chow hall) to eat. I honestly didn't mind the job. I forced myself through intense workouts preparing for boot camp and felt I was gradually losing my conditioning over the accumulating weeks. Being a crossing guard was actually beneficial in that regard. The biggest disadvantage came when marching to eat a meal. We, as a group, were on a timer in the cafeteria. The clock started with the first recruit of our company of men entered the building. I was one of the last to get food; therefore, I had the least amount of time to eat it.

Entering the cafeteria, we walked single-file, with a tray, and food was placed upon it as we passed by. We then proceeded to large round tables to shovel down the food. Time was ticking. By the time I sat, I had perhaps five minutes to eat all I could. Then it was back outside to assemble and march to our next destination. I still blame that experience as my reason for eating way too fast.

As time passed, our COs yelled less and interacted more, but they were never congenial. As I said earlier, there seemed to me to be an obvious outline they were following. In keeping with my original instincts, I followed orders and stayed in the background. That plan resulted in never receiving the face to-face tongue-lashing some received.

However, during one inspection I had a brief encounter with the shorter CO. He and I were the same height and so we were eye-to-eye. He noted a single black chest hair extending from beneath my white t-shirt. He quietly informed me of this observation, telling me he could resolve this problem if I had no objection. I said nothing, simply maintaining my focus straight ahead as I stood at attention. He reached up and plucked it out; thereby; resolving the issue. My jaw clenched, but I didn't utter a sound. He then moved on. Honestly, I would have done the same had I noticed it. I thought, if I get through boot camp with that as my biggest problem, I'd be happy.

I was naive – that didn't happen. In time I exposed my weakness and paid for it.

Some weeks now had passed and we were encouraged (I refrain from saying it, but we were ordered) to write home. They provided us with pen, paper, and envelopes. I did so happily. It provided some calm during a hectic and anxious time. My first letter was to my parents summarizing my experiences. I don't remember specifics in the letter. My hope was to relieve any worry they might have. I then penned a second letter, this time to Ellen.

I thought about the last evening we spent together, walking and talking. I sat and wondered if she had thought of me at all. I hoped she had. We never dated and I mentioned earlier it's really not accurate to say we had a relationship, but I'd been wondering about her and what my return to see her would be like. Certainly there is a part of me that misses her and wonders if we would have seen each other more if I hadn't joined the Navy. I placed both letters in the mailbox and excitedly hoped for letters in return.

Boot camp is more than halfway over and I am excitedly looking towards its end. I'm ready to move on. It was announced one evening that we could test for the Navy Seals if interested. It amounted to an easy workout by Seal standards, but not for me. I was not interested in becoming a Seal but signed-up with the thought of it breaking the boot camp routine. It was a way for an evening out of the barracks.

There was no way I'd abandon the goal of serving and experiencing life aboard a submarine. That and the electrical/

electronic training I'd receive was my ticket to a rewarding career after the Navy. I was still devoted to that plan. I felt in control of my destiny now. Serving as a Seal was certainly admirable, but not my future, but I did the Seal testing that was offered. I repeatedly did pushups, sit-ups, swimming, and treading water along with other things both in and out of the pool. After some time, I checked out. There were no questions or shame. I returned to my barracks and resumed the boot camp routine.

The monotony of classes and testing continued as did the nights without restful sleep, repeated inspections, and rushed meals, but the end date is in sight and it can't come soon enough. I feel the need for change again. The boot camp rules seem sometimes without merit. I know their purpose is to build us in the Navy way so as to become reliable sailors, but I feel I've gotten the point and reached their goal. The evenings before bed are now our own to relax.

A few recruits somehow noticed the commanding officer's office was left unlocked and a couple candy bars left in full view on a desk. I love chocolate, especially dark chocolate, but would treasure any type of sweet treat at this point. Apparently, several other recruits felt the same. If this was a trap, it exposed my weakness perfectly. We all split the candy and ate it, almost swallowing it whole. I think that's the only way I know to eat now. I knew I'd get in trouble before the first bite but did it anyway. I don't remember the brand of candy except that it was fantastic!

The next day started with the COs noticing the missing candy. My three friends and I admitted to the crime. Surprisingly, there was no yelling, but rather they made us aware that the punishment would be "... an evening of hell". That entailed pushups, sit-ups, running, and other calisthenics to exhaustion. I got through it by focusing on performing better than the others. I admit I could have done without the experience. It lasted for more than an hour and a half, maybe even two hours, and then we were sent back to our barracks. It wasn't fun, but I did it to myself. I couldn't be upset – it wasn't unfair.

Boot camp graduation was now just a week away when I received a letter from her. By this point I had sent and received a few letters from family. Receiving mail was uplifting. It was nice to know I wasn't forgotten and perhaps missed a little, but Ellen was who I really wanted to exchange letters with and eventually see again.

Looking back now, I don't remember who actually wrote the letter to me. It said something like she would be given my letter when they could, but it might be some time before they are able. She was doing well and I was thanked for the letter. This response really caught me off-guard. Obviously they had opened and read my letter. I didn't know what to think or what I should do, if anything. At this point I didn't know if I'd get time at home after boot camp or go straight to submarine school.

I wondered what possible circumstance(s) this letter could be referring to. If I did get a break home after boot camp, would this situation keep us from seeing each other? I didn't write a return letter. I would wait to get my orders and hoped I could develop a plan from there. There was no one to talk to even if I weren't in boot camp. I never met anyone in her circle of friends or family and never expressed my thoughts about her to anyone and now perhaps never will, but I was concerned for her.

The day for graduation was here and I couldn't be more excited. I was honored with being selected as Honorman for Company 008. That's important for boot camp and my Naval resume, but beyond that no one cared. As a side note, only the Honorman received a name badge for our uniform. My parents were impressed and proud, just like at college graduation. They have always been supportive, no matter the circumstances.

Prior to this ceremony, everyone had their picture taken in

uniform. Mine was with the Navy-issued, black-rimmed glasses that in my mind were hideous. Fellow recruits referred to them as birth control glasses since no girl would be caught dead with anyone wearing them. At least that was their reputation, but my company commander didn't allow me to remove them for the picture. I never sought a copy of that photo. I hope no one has it.

I was also given the rank of E-3 upon graduation. That was stipulated in the program that I will be attending and the training I'll receive. Following some training yet to come, an automatic advancement to E-4 would be next. So by the time I step foot on a submarine, I'll be an E-4 – almost halfway up the food chain. The enlisted ranks run from E-1 to E-9 as follows:

E-1: Seaman recruit
E-2: Seaman apprentice
E-3: Seaman
E-4: Petty officer 3rd class
E-5: Petty officer 2nd class
E-6: Petty officer 1st class
E-7: Chief petty officer
E-8: Senior chief petty officer
E-9: Master chief petty officer

This is the first step in my Naval career, but it is as good or better than I planned or knew might happen. As long as I'm not enticed with more chocolate, I think I'll do well. My orders following boot camp are to proceed directly to submarine school in Groton, Connecticut. The weather at boot camp graduation was sunny, with a slight cold breeze resulting in a small wind chill, but no clouds to threaten rain. I'm accustomed to this weather by now.

My parents and I had a nice visit together. Since I had to proceed directly to sub school, we exchanged hugs for hello and goodbye in the same visit.

So a visit home is delayed. My thoughts of Ellen remain. I hope whatever circumstance or whoever is keeping my letter from her is also keeping her safe.

Losing someone we kept in our heart
is a loneliness concealed, but never to end.

Chapter 6
Sub School

I wasn't alone travelling to submarine school, but we made up a small percentage of those in boot camp. If memory serves, there were two in addition to myself from the eighty-man lot. Having been awarded the rank of E-3 out of boot camp, I was the highest ranking of our three-man contingent. Arriving in Groton, Connecticut meant becoming a part of the regular Navy world (admittedly to a far lesser degree). Not yet being submarine qualified and thereby not having earned our dolphins meant we had much work to do. We were only of limited use to the Navy until qualified, but still, the path we're on is exciting.

We checked in and were guided to our barracks, shown the chow hall and where to report the next day to receive additional instructions. No marching was required/expected and no one ordering us to attention anymore – also no yelling. The petty officer checking us in did take note of the rank I received and the recognition of honorman. I didn't verbalize it but was appreciative.

Walking around the base was amazing! I'm feeling good about my life's direction. Between buildings I got a small glimpse of the submarines in port and can imagine myself getting on board, but also wondering about what daily life is like on board. I feel the adrenaline pumping. This is just a submarine base – no surface ships. The terrain of the base includes hills. As I look around, I see non-descript buildings, fencing, guard shacks, and what I'm most excited about, the submarines. My view of those are unfortunately limited. The weather is mostly grey clouds though no rain and the ground is dry. No wind is present, but I'm wondering if there is ever much wind due to the many buildings on base and apparent forest surrounding it acting as a wind block. Walking around required a light jacket. The grey skies and coolish temperatures reminded me of an early fall day in southern Ohio. While I know myself to be early in my training, I feel I belong, albeit all alone.

During boot camp the Navy preformed their due diligence in granting me the initial security clearance required for my job. I have no idea who or if neighbors, friends and former employers were contacted, but potentially they could have been. No problems were encountered as far as I know so things have continued smoothly.

Sub school is scheduled to be done quickly (approximately 6-8 weeks) just like boot camp. There is nothing the Navy needs from me on this first day so my thoughts drifted to the girl I left behind – the girl who hasn't been able to receive my letter. I find myself distracted by her so decided to write again. Besides, my return address has changed from boot camp to sub school so I want her to know that, just in case. Maybe she'll get both of my letters soon and we can plan a get-together after sub school. That's the hope I cling to. Writing now also gives whoever is helping her or perhaps even herself many weeks to respond. For now, that's all I can do. Starting tomorrow is sub school and it demands my full attention.

And then there's the girl six years his junior in a hometown four hours north…

More time has passed working her two jobs. And living at Grandma's remains comfortable and easy. Nothing has changed except for her loss of patience with her life and career. It's becoming an ever increasing problem so she recently forced a change.

In a week she's starting a new job at a local drug store. She has considered this change for a while and is excited to be leaving both of her current positions. It's incredible that she maintained a lifestyle of two jobs for so long. Styling hair isn't something she aspires to anymore, it was just her best option at the moment. She can't envision a long-term future either professionally or personally. That thought saddens her, but she disregards it for now.

For the first time in years, she will have just one job. She feels a sense of freedom because of that and a reason to be happy. Aspirations for the future are being delayed for a while. Some sort of vacation in the near future is still her hope. She'll be able to afford it by summer. Hopefully, she and her girlfriends can figure something out, but unfortunately, her personal situation with the boyfriend might also need changing. She won't press him or their situation just yet. That would result in too much change.

One thing at a time, but this might be the year everything changes.

Dan's training continues…

Sub school had some interesting diversions, but it predominantly consisted of classroom instruction. We were a group of twenty-five students. Knowing how to take notes was important since most of the reading and instructional materials were not permitted to leave the classroom. They contained classified material so they were accounted for and locked up at each day's end. Each of us took turns in that accountability each day.

The subject matter wasn't difficult. The challenge was that we were being given a lot of information to absorb quickly and it was truly foreign to all of us. There really isn't anything in ordinary life that compares to the submarine experience. I would later learn the instruction I received was particularly relevant to me since the class and type of sub we were learning would be the same class of submarine I would eventually be assigned to. That was not true for everyone.

The start of each class was us listening for our instructor to arrive. I know virtually nothing of motorcycles–but his was a Harley-Davidson. It was black, a larger model with saddlebags, handlebars set to a height requiring his arms to be extended straight out and slightly above his shoulders as well as tassels dangling from the ends of the handlebars. Everyone knew when he arrived because of the loud rumble of his bike. He wore chaps, black leather boots, leather jacket, and a headband under his helmet. He once said he wore a helmet on base because it was required but removed it as soon as possible upon exiting the base at the end of his day.

The helmet he chose was matte black with a chin strap, but no face shield. Looking out the window from our classroom, everyone took notice and responded with an incredible and spontaneous outburst of laughter. He looked fairly ridiculous, at least to all of us. Thank goodness neither he nor other instructors witnessed our reaction.

He parked his bike and we took our seats. Some minutes later he strutted into the classroom in uniform but wearing a scowl. Was he angry? It was an ambiguous first impression. This guy was hilarious, but not in a fun way. He seemed to think himself impressive.

The classroom experience was similar to high school or college. We took notes during lectures and referred to books, videos, and handouts we kept. The expectation was also similar in that we were to study at home (our barracks), attend study sessions, and seek individual attention as needed to address questions, but as previously mentioned, a few diversions were like nothing I've ever experienced. Most memorable was the pressure tank, flooding simulator, and fire-fighting trainer.

The pressure tank's purpose was to simulate the effects at one hundred feet of submerged depth. A small group of us entered and took a seat in the tank with an instructor. Another instructor waited outside the tank. We were instructed to 'pop' our ears frequently by holding our nose and mouth shut and gently blow. Just like what many do in an airplane. The simulated depth was increased with occasional breaks to ensure everyone was tolerating the experience well. I could pop my ears effectively, but a couple of men experienced slight bleeding from the ears due to their eardrums bursting. We were warned this could occur, but also told it would heal without significant intervention. The whole experience was interesting, but I can't say I felt any different at 100 feet. Still, I was happy we did it. We did wonder if we would sense pressure when submerging in a submarine. This is jumping ahead, but I ultimately never sensed any effects from increasing depth during any patrols on a submarine.

Things are happening at home. My college roommate is getting married and I was able to get permission to fly home, but just for the weekend so unfortunately I couldn't stay long enough to do anything but enjoy the ceremony and quickly leave. That was okay, because it was the only reason I want to go. The Wednesday before my Saturday morning flight, my instructor told me I might not be able to go. I nearly broke into a panic. I was quite angry with him announcing this so close to the weekend. The cost of the airfare wasn't refundable. He didn't provide a reason for cancelling my trip–but I didn't ask either. I did ask if there was anything I could do to regain the permission I had.

He gave me extra duty to complete by the end of class on Friday. It wasn't hard, I just had to satisfy his compulsion to show he was in charge. With his ego fed he gave me permission, again. The wedding and reception were fantastic. I reconnected with friends I'd been missing and bragged a little about what I was going through in the Navy. And thankfully, airlines ran on time.

Returning to base from the weekend reprieve meant experiencing the flooding simulator. It is an adrenaline pumping and definitely appropriate learning effort as it is certainly must-know skills for submarine duty. Obviously, flooding in a submarine is very serious. As I'll talk about in a future chapter, submarine qualification requires memorizing valve line-ups to stop flooding from a ruptured pipe, minimize damage, isolate the compartment and ultimately evacuate the affected areas. We participated in this exercise more than once. It was important to know how to use the tools as well as how to stop the various sources of flooding – be it a pipe, flange, valve, or some other fitting.

The simulator is basically a very large water-tight tank.

It was stacked with tanks, piping of various diameters snaking through the area along with flanges and different types of valves. We'd practice stopping a simple leak to high pressure ruptures in piping or flanges both above and below the water line in the flooded compartment. Sometimes briefly holding our breath under water to perform repairs, instructors gave us very limited time to accomplish this for safety reasons. They only let the flooding get to waist deep. I loved this trainer. In one session a small cut on my forearm was significantly worsened when hit by a jet of water. There was no pain (probably due to adrenaline) but was pretty bloody. It looked worse than it was.

After each training session we'd get a debriefing to review what was done correctly and what/how to improve. This trainer was incredibly fun to be part of the team, both in the simulator or watching at the observation window.

Sub school is progressing at a fast pace. It's now halfway over and I'm anticipating its end. The men in class will go in different directions afterwards since some of us will serve aboard fast-attack submarines and some on ballistic missile subs. I suppose we'll all eventually believe our type of sub is best, but for now we're one proud group. I won't actually get on a sub for probably another year, some will go directly there after graduating, but we all share the same excitement.

Perhaps only a week or so later was fire-fighting. This was also a blast. In groups of two we took turns extinguishing a fully engulfed compartment. One person at the end of the hose operating the valve and controlling the direction of the spray. The second person behind him holding the hose and observing for safety concerns. We each took turns at the two positions.

The fire trainer itself was a relatively small compartment with simulated equipment and piping. Attire was fireproof clothing, helmet and an OBA (oxygen breathing apparatus). Knowing how to don an OBA, properly fit the mask, load/activate the canister, and eject the canister that created the oxygen we breathed was very important training. We practiced that on its own several times both at sub school and eventually as part of the submarine crew.

The fire in the trainer was lit. Flames quickly filled the room, crawled along the floor and scaled the walls. I manned the valve as the front person and a fellow student behind me as an instructor charges the hose, meaning he turned on the water. The large diameter hose filled immediately and in doing so attempted to jump from our grasp. We approached the fire slowly in a deep squat, one step at a time.

I glanced at my partner to indicate I was about to open the valve. Upon doing so, we got the expected jerk backwards. Moving ahead I swept the water spray back and forth, left to right, at the base of the fire. My partner ensured we didn't inadvertently walk under flames or get encircled. As with the flooding simulator, adrenaline was pumping. Our debriefing went well and then others took their turn, but as fun as this experience was, I hope to never do this on a submarine.

As a group we all enjoyed those "distractions" of hands-on training. We talked about it for days afterwards. A few of us of decided to celebrate a little with a trip to Boston, approximately a two-hour drive if memory serves. I don't know why, but thankfully one of the guys drove to sub school so we had transportation. Rental car cost would have doomed the trip. None of us were making much money.

Being idiots, the four of us guys were caught car dancing to Cindy Lauper's *Girls Just Want To Have Fun.* Making matters worse, we were in full dress uniforms. It was at a red light. A car stopped next to us and got one hell of a laugh. We laughed as well. Our embarrassment didn't stop the celebration though.

Later, we accidentally ran a red light in Boston because we honestly didn't see the traffic light. Traffic lights were sometimes positioned along the right side of the street and sometimes overhead. Thankfully, the officer pulling us over understood and gave only a verbal warning. Perhaps being in uniform helped since we were obviously from out of town. We spent just the one night in Boston and saw nothing, but the inside of a couple taverns. Other patrons bought all our drinks. Being in uniform made that happen I think. We had the best time there. Someday I'd like to visit Boston in a more adult fashion and see everything it has to offer.

Back at base classwork continued to go well. My grades were quite good and the weeks passed quickly. The base also had a submarine museum that I regrettably never visited. I've always intended to somehow get back to see it.

Sub school was nearing an end. We planned a picnic to celebrate as a group. We weren't permitted alcohol and didn't want our instructor to attend, but he did and was mostly ignored. My feeling was that he didn't care for that treatment – I was pleased with that thought.

My next duty station will be Dam Neck, Virginia for electrical/electronic education and then finally my job training, but first I was granted a week at home before being required to report. I was happy to visit family and relax a little. I hadn't received the letter I was hoping for. I suppose Ellen either can't or is not interested in responding. I had such high hopes.

I was flying home from sub school, leaving Connecticut behind, thoroughly intending to enjoy the next week. The previous four months had been a whirlwind. I wanted a change and certainly got it. I hadn't made any plans in advance except for my parents to pick me up at the airport upon my arrival. There's plenty of people to call, though their work schedule might present a difficult work-around, but I'd be satisfied if all I did was relax.

My parents were happy to see me and I them. The spring weather was decent, the sun bright and no rain in the forecast. I was happy–finally. Honestly, it was a nice change to be at home and not hopeless for the future. The rest of my first day home we enjoyed each other's company. I explained to them what was next for me and in typical fashion, my Mom implored me to be safe. Dad listened and smiled. Mom made dinner and they updated me on family comings and goings, so to speak. I slept well that night. I had forgotten how comfortable mattresses could be.

After a slow next morning and lazy afternoon, I reconnected with a friend. We met at a familiar bar for a drink. I think we were both glad to see each other. We had so much to talk about. Just as with me, his life had many changes. He was now a businessman with a steady girlfriend and little spare time. Spending a couple hours with me was an exception to a busy routine and I was very appreciative. I'm glad to see how happy he is now. It's hard to believe at one point we were both considering the Peace Corps. The evening passed quickly with me soon to be back home in front of a television with my Dad.

The next couple of days I spent with family – my brother, sister, their spouses and kids. Nice, casual gatherings enjoying each other's company. There was downtime though when my thoughts wondered toward Ellen. I don't know when I'll next be home again.

I decided to go to the grocery store across town with the hopes of seeing her and maybe understand how she's doing. I hope me coming to see her isn't upsetting. I thought of the possibility she chose not to write and wasn't interested in seeing me.

I borrowed my parent's car and drove to hopefully see Ellen working at the grocery store. It was around lunchtime. I hoped to catch her taking a break and talk a while, but unfortunately she wasn't there and the couple of staff members I asked had no idea where she's been. She hadn't been there in a while. Maybe she got a new job, they speculated. I suppose I'll never know. She really didn't have any obligation to tell me so I suppose I need to accept it and move on. I hadn't met her family or friends so was very limited as to who I could ask.

Still, I can't help wondering why my letters were held-up. It would be nice if I could figure out a way to talk to her. I returned home to my Mom passing along a stack of accumulated junk mail to me. However, one letter was different than the rest – finally the letter I hoped to get!

If you really want to know what I'm thinking when I cry,
Be prepared to lend an ear while I spell out a lie.

Chapter 7

New Base, New Training

The weather was perfect upon my arrival in Dam Neck, Virginia. The Navy base is only a short drive from Virginia Beach; the time of year was May or June, I believe. Sun was bright and the temperature perfect with only a t-shirt – a spectacular summer day! A spattering of very high thin clouds moved past as I looked up. Flowers were in bloom at the airport to welcome visitors and the grass and trees green. A very easy breeze wandered about and felt wonderful. I felt welcomed and looked forward to getting started. The training was to be my last stop before getting on board a sub, but I'd have to be patient – unfortunately not a strong trait for me.

This training is scheduled to last one year, but this education is partly why I agreed to a six year (versus a four year) enlistment. The electricity and electronic skills learned here are directly applicable to civilian jobs.

Naturally, the barracks are clean and well maintained since the students are held accountable for that and regular inspections verify it. They are two story painted cement block buildings with commercial grade vinyl tile floors throughout. Two men per room with one common bathroom per hall that included showers. The rooms had a bed, desk, and free-standing closet for each person. The base was small enough to require just a short walk to the chow hall, the Navy Exchange for uniform, and classroom supplies and such, or to the beach on base.

Even though some sailors were technically underage in some states (less than 21), everyone was permitted to have and drink alcohol at the beach on base – that wasn't true on the public beaches. I thought the food in the chow hall was fine, though some complained, but a few 'complainers' always seem to be the case. The next years' worth of training was divided into "A" and "C" schools with A school preceding C school. For me, "A" school was electrical/electronic training. "C" school for me was Fire Control Technician-Ballistic training. I was to be a FTB. "C" school was specific to the job assignment on board a submarine. A number of different job training "A" and "C" schools were at Dam Neck. We were expected to keep our rooms spotless, with beds made, Monday through Friday. There was a rotation among us to also clean the rest of the building on weekends, but that was minimal and easy work.

Each hall of the barracks assigned one person to attend a monthly meeting with the commanding officer in one of the lecture halls. For a while, it was me. The purpose was to provide feedback or ask questions and for him to communicate directly to us; thereby, avoiding the confusion that sometimes occurs when word is passed down the chain of command and among sailors. I honestly kept pretty quiet. I was relatively new and not yet comfortable speaking out. I did it until another new guy arrived.

"A" school was more classroom training, but with frequent labs or hands-on lessons. Roughly, our hours were 9a-5p, Monday through Friday. Study sessions were in the evenings after dinner though not always compulsory. Additional duty on a rotating basis was manning the desk at the entrance of the building checking IDs as sailors arrived for the study sessions.

Besides occasional barracks inspections we had personnel inspections. All instructors were Navy personnel and, I thought, quite good. I found the classes interesting and their sea stories enjoyable.

Summers with access to Virginia Beach or the beach on base spoiled us. A few drinks and days on the beach soaking in the sun on weekends was our routine.

One particular day began in the usual way. The usual breakfast, the usual banter and the usual plans. The three of us made the usual plans that would fill the typical weekend day. Each of us were young and without goals we were interested in discussing or plans that extended any further than dinner that day. The youngest, Clay, was like a leaf in a stream, as the saying goes. He would easily move in whatever direction he was encouraged to go. He didn't consider consequences or explore risks. He was good at doing what he was told. Mitch was entrenched in repetition as a means to live life without disruption. His hope was for each day to be the same. Habit kept him safe and not noticeable. Both were in training to become Missile Technicians (MT) because that job required (and they are good at) working with their hands.

The oldest (me) and most experienced of the trio provided a perceived leadership and a guidance without risk. Neither the leadership nor guidance were actually true. It was now late morning and the plan was set. "Let's do what we always do." We spent a lot of time together when not in class and that time was often at a beach or in the town.

On base was something very unexpected for me-lunch every Friday. Instead of going to the chow hall, some men ate at the E-Club (enlisted club). That was because of the dancers. I found it unbelievable this was allowed. The women wore outfits that were basically bikinis. Nothing less than that.

I attended a couple times with friends, but sat in the back, far from the dance stage. I was ultimately too cheap to pay for lunch when I could get it free at the chow hall. Plus, I honestly didn't find the shows that captivating and had no interest in placing dollar bills on the dancers. I needed every dollar given me by the Navy. I wasn't good at financial decision-making, but knew those dollars weren't a good investment.

Ultimately, the dancing stopped when a female commanding officer took over. It was a good and long overdue decision. I often wondered what the female sailors thought. Besides, some classmates occasionally drank slightly too much to make the afternoon's classes beneficial.

Going to the beach had become just a place we frequented, but we were boring and drab and so were our outings. Looking back, that's obvious more than ever. It wasn't our fault entirely. Virtually everything we did, wore, and ate was determined by others (the Navy), but honestly, also exacerbated by our lack of creativity. The life of nearly everyone we associated with was also drab, it was the Navy way. The weekends gave us opportunities to live a little I suppose. We walked a little taller with chests a little more forward and a confidence a little more buoyant (no pun intended).

I'm sure passers-by looked away due to feeling embarrassment for us. Beach attire was Navy-issued and muted as if it had been worn and washed for years. Hair was so short, sunscreen applied to the scalp was a serious consideration. None of us participated in any additional exercise regimen so our only "bulking up" occurred in the waist from beer.

This featureless group travelled in Clay's equally non-descript vehicle. It was a large four-door grey sedan with rust dotting the lower portions and required way too much gas to produce movement. For whatever reason, our history of meeting women wasn't good, but then I personally wasn't ready yet for anything that could be defined as a relationship. Ellen was not distant in memory yet. This short trip to the beach, it turns out, was the ultimate proof of the power of attraction.

In school I studied hard. I wanted to do well and not waste the time like college. I really felt the Navy was my opportunity of a lifetime. As I mentioned previously, my life's direction would be determined by my Naval career. One enjoyable distraction was answering the written questions from a grade school class back home. A friend and school counselor back home asked if I'd be willing to do so. She organized it and sent the stack of letters created by the kids in a manila envelope. I wrote answers in basic and general terms. I honestly didn't know how detailed I should be for that grade level, but also couldn't provide detailed answers to some questions due to confidentiality, but I enjoyed the challenge and hoped they enjoyed receiving my answers. It was a fun exercise. Many years would pass before I re-connected with that school counselor. I still wonder how my responses were received.

At the beach were three women who had arrived as tourists, having survived a considerable drive from a northern city. As it turned out, they came from a town only four hours north of my hometown. They were there to escape, forget, relax, and release some tension and emotional baggage, especially since one had a recent break-up. They had purpose, but limited time. The days for this escape were numbered. They had a week, but that included the days necessary to drive there and back again.

Hurriedly, they established their small motel room. The organization of their suitcases was disrupted by the search for their bathing suits. Everything not immediately needed at the beach was left for later. A frenzied, but determined effort is an apt description of their beach preparations. With small bags packed, and after a brief walk, a small expanse of beach lay before them. They gathered briefly to select the best spot. Then with a focus resultant of weeks, if not months, of waiting, the women were comfortably established.

The three were quite different. One was quiet, difficult to approach, and required much effort to engage. She would require the kind of attention that only comes with immediate and overwhelming infatuation. Another of the women was nice and outgoing. Someone anyone would be glad to call a friend. Being more than a friend though would entail much empathy regarding the sporadic bouts of neediness. And lastly was the woman who provided the link for all three. She had cut ties with her boyfriend after a considerable time dating and thus was one reason for this escape. She was smart and quiet. She was a hair stylist by training but wasn't currently in practice. Instead, she has been working at a local drug store. She had saved her money for quite a while to afford this trip. This small company of vacationers were beginning to relax. The beach would fill their days, but the nights were not yet determined.

The three of us bland looking men parked our bland car arbitrarily. We had only a short walk to the beach, exhibited smiles without reason and a swagger well-rehearsed. Upon reaching the boardwalk we paused. After only a very short assessment, one of us spied those three women. Mitch unexpectedly had excitement that was sudden and seemingly without bounds.

He pointed, "Look, look, look!"

Clay and Mitch could not contain their excitement. I (the oldest) just wanted to relax on the beach – I was in no mood for the necessary small talk involved with trying to meet and impress anyone. Plus, any thought of meeting a girl reminded me of Ellen, but I held the minority opinion so toward the women we strolled.

The women were sitting in a row, each on a beach towel facing the ocean. This is only important because I was selected as the one to take a place next to one of the girls on the end. My two friends took places next to me, creating a line of six people facing the waves. It was an odd site from a distance and uncomfortable up close.

As I had intended and made clear prior to taking this position on the beach, I laid down, closed my eyes and tried to relax. Nothing in me wanted to take a chance on anything becoming a relationship of any kind.

Obvious was the intentions of two out of three of us though as there was an endless expanse of beach we neglected to be almost uncomfortably close to these three girls.

Some time elapsed until the girl next to me was apparently selected (or perhaps obliged because of our un-necessary close proximity) as the one to initiate conversation. I heard her but chose to ignore the friendly overture. Luckily my friends did not. A fair amount of conversation entailed. I sporadically volunteered comments. Let's face it, I was rude. The afternoon passed quickly. At some point they all settled on a plan to meet later at the girl's motel room. No one really cared if I attended, but I did.

The three of us guys were showered and dressed for the night. We each applied too much cologne and raised our expectations beyond what's reasonable. All of which was standard practice-which is probably why dates were rare.

I found myself excited for the evening even though I promised myself I wouldn't be. We arrived with suitable drinks for the group and were met by an equally impressive purchase the ladies had made. Us sailors imbibed quickly and repeatedly. It was necessary so as to gather back the confidence we lost after leaving the beach. The initial conversations were to figure each other out. There were the usual questions. It turned out we were all the same type of people with similar backgrounds; all from Midwest middle-class families. The evening was moving quickly. The women also imbibed considerably. We laughed and talked in pairs, small groups, and all together. We spent time between the beach and the motel room and motel room balcony. What a fantastic night!

The girl with the recent break-up and I hit it off eventually. I stress it occurred eventually because the first impression of me was understandably bad. Looking back, I don't remember how the night ended – that is except for a brief two-sentence verbal exchange she and I had.

In my memory, the night moves as a blur, but not due to our beverages of choice as the causal agent, but because of the two sentences we exchanged.

Some months later the two men who accompanied me that night finished their naval training and moved on to new assignments. I never saw or talked to them again. It's unfortunate that those two people so influential in my life's direction have virtually disappeared. Although I didn't really know the other two women, I also never saw them again either. But that one lady among them, the one I inexcusably ignored initially, turned out to be my destiny. You may wonder as to the two-sentence exchange.

I asked, "Do you want to get married in about a year?"

She responded, "Okay."

It is perhaps ironic that she and I had training in specific fields that we both disliked and didn't wish to pursue. Hers was hair styling and mine was paper technology. She was six years my junior by age, but older as measured by maturity.

She was beautiful and smart (and considerate, humble, and caring) and yet she chose me. I promised myself this wouldn't happen. I told myself I wasn't interested in relationships. I convinced myself I'd only get hurt again, but she was different from anyone I had ever met. Somehow, she affects my soul.

Time together and now a new friend,
Refusing to dream is gladly at an end.

Transition, “A” to “C” School

I've been in the Navy for around eight to ten months to this point. It'll be about six to eight more months in the classroom before knowing my submarine assignment. Classes and lab work are increasingly more difficult. We are now able to understand and troubleshoot electrical diagrams. “A” school will last another two months or so. Then the “A” school graduates will separate to start their selected “C” schools.

Our first C school instructor will be Petty Officer Henderson. His first name is Bob, but the formality of the Navy prevents me from calling him that. He has a wry smile that to my eyes indicate a devious and fun-loving attitude. He smiles when introducing himself and looks the other person in the eyes when conversing. I think I'm going to like him. I feel like he has a past I want to hear about.

I was able to buy a cheap car. It was my first experience buying and actually owning a car. My excuse for the car I purchased was that I was in a hurry and terrible at negotiating. I also had very little as a down payment and a rather small income from the Navy. I got what I could afford from what was available – a very unimpressive, small, silver, used two-door Dodge Colt sedan. It was an older model and had a little rust dotting areas below the doors. It was quite a small vehicle. I'm not tall but could sit in the driver's seat and reach across to manually open the passenger side window. It had black vinyl seats and very little suspension for support. The interior black color was hot and sticky in the southern heat.

I and two other students decided to move off base to an apartment in Virginia Beach. The apartment we selected was a three-bedroom unit, but we had to walk through one bedroom to get to another which was kind of inconvenient. It had a small porch surrounding the front door that we hardly used. Instead, we came and left through a side door that had us enter directly into the small kitchen with an extremely small dining area. From the kitchen one would pass through another door to the living room.

Each of the three of us were starting a different "C" school. We continued to party on weekends, especially since we no longer had the regiments associated with the barracks cleaning. Our large living room window facing the street was the perfect place for our accomplishment-the display of a pyramid of empty beer cans.

Pete was from New Jersey. He was a very likeable person, easy to meet and talk to. A really nice guy. He planned to become a Missile Technician (MT). Greg was the tallest of us three, blond and gregarious. He could be something of a loose cannon, but also very nice. He was training to be an Electronics Technician (ET). We all were similar in our outlook of taking each day one at a time. None of us looked to the future in any very specific ways, all of us just very casual in our approaches.

The Navy gave many sailors the independence they sought before they were mature enough to handle it, but I suppose college does that too. I heard of one weekend outing that to this day sounds somewhat shocking, but unfortunately believable. Drunk sailors out on the town weren't unusual, but I never knew of any fights or problems. As the story goes on one particular celebratory night, two drunk sailors put a third on a non-stop bus heading to a town two hours west. He apparently slept the entire bus ride until forced to exit by bus personnel.

Cell phones didn't exist yet. He had no idea where he was

when he got there. He later tells the story of bus terminal personnel having strange facial expressions when he asked what city he was in. His rest in the bus terminal was sporadic and the night harrowing. Luckily he had the cash to buy a return trip the following day. There are so many stories that could be told. The level of truth in the stories should be suspect.

"C" school started with a couple changes. I increased in rank from E-3 to E-4 (petty officer third class) and the classes shrunk to approximately sixteen men. After being in the Navy for over a year I finally was receiving the instruction for the job I'd perform on the submarine. The training was specific to the submarine ballistic missile fire control system and missile guidance system. This gave me a purpose. I was learning something that civilians have a hard time relating to since their exposure to it is mostly through movies. My job in the Navy and serving on a sub would be unique and full of experiences that civilian life can't offer. That was a goal for me and one I'm on track to attain.

Our apartment off base was no place anyone would call home. No one making a decent wage would have considered this place. The military wage left us rather poor, quite honestly. Although, we also weren't wise with the limited disposable income available. Greg and I would occasionally race home from the base for no reason at all and for no prize. We were in different "C" schools and sometimes found ourselves finishing our day of instruction at the same time. That meant we scrambled to our cars and sped to the apartment. We drove too fast, swerved too often, and might have run an occasional traffic light in its transition from yellow to red.

His car was definitely quicker than mine, but road congestion and traffic lights were the equalizer. I don't remember the totals, but felt I won often enough. Winning the race meant nothing except having that honor until the next race. That was fun and irresponsible. And I don't regret that fun.

The apartment itself was quite sturdy with walls made of concrete. As the three of us would finish our beers, we'd throw the cans into the family room wall to see who could produce the most dented can. The walls were fine – no damage. It probably should be embarrassing that we found that entertaining.

Winter was quite uncomfortable. We often left our coats on even when relaxing inside. The apartment was heated by a small heater in the middle of the apartment's living room. It was three feet long and wide and about two feet tall. That meant the three bedrooms and kitchen never warmed – only the living room. Drinks left out overnight in the kitchen during winter froze.

We built a contraption out of metal hangers and aluminum foil to deflect the heated air exiting the blower towards the rest of our apartment, especially our bedrooms, but it failed to produce any noticeable benefit. In spite of all the negatives regarding this rental, it was only three blocks from the beach. That allowed us to look past a lot of the deficiencies.

So far, boot camp, sub school and "A" school have gone well, and "C" school is continuing the trend. I'm proud to say I excelled, placing first in my class. That meant I had first choice of the various openings on submarines. The C school staff decided that the available submarine openings wouldn't be assigned, but rather students would choose in order of class standing. I chose a sub with only one available spot. That way no one from class could accompany me. There were a few students I wanted to leave behind. Following graduation, a couple of the men rented a beach house and threw a party. We all attended and had a blast. Most drank too much, but having the home meant no one had to drive afterwards.

My car was really no better than the apartment. As mentioned, it was a small silver Dodge Colt. The body-side molding had fallen off. I stored it in the trunk. The radio turned itself on and off spontaneously with bumps in the road. It was small enough that I, as a relatively short guy, could easily reach anywhere in the back seat. I didn't know it at the time, but in a year I'd be taking it to the junk yard because the dealership refused it as a trade-in. The structure surrounding the two front struts was so severely rusted, it appeared the hood was the only thing holding them in place.

Not even the junkyard rewarded me much for it.

My final obligation before we all separated to our chosen duty stations and submarine was to attend a picnic with our instructors to celebrate our class graduation. Petty Officer Henderson, or Bob, came dressed in sharp, but casual attire that his wife helped him select. He wore a light-weight button-down shirt with a pressed collar, white shorts with a crease down the front, and sandals.

It was a great weather day. The sky was clear, no clouds, only bright blue above us. The temperature was in the upper 70s. No breeze though, so shade was helpful to get breaks from the sun. We had purchased bottled beer, ice for coolers, bags of chips, and burgers for lunch. I was part of the early arriving group to the park to reserve a grill. I and another sailor opened a beer and loaded the grill with charcoal and saturated it with lighter fluid. We applied a lit match and excitedly watched the burst of flames leap high above the grill. We were obviously too liberal with the lighter fluid.

As the briquettes gradually turned grey, people arrived. We drank until it was time to cook. I manned the grill. The burgers cooked fast and as the grease dripped onto the coals, small flames would briefly appear and smoke billow out. I waved the spatula to blow away the copious amount of smoke created by the burning grease. Bob was next me and we had a long conversation. I wished I could have served on a sub with him. He had some great stories.

Eventually I called for a plate to transfer the burgers and let everyone eat. Bob and I looked down as I scooped up the burgers and both noticed it at the same time. The waving action by me with the spatula caused grease to splatter and dot his white shorts. They looked awful. He said his wife had just bought those shorts for him to attend this event. I wanted to burst out laughing, but held it in. Those shorts didn't look salvageable to me and I bet he was going to be in big trouble when he got home. We decided to chug the rest of our beer and open another. That's how I remember Bob now. I've forgotten all his stories. His memory of me is probably just the guy that stupidly ruined his shorts.

My time at Dam Neck, Virginia was almost over. I'd been in the Navy for one and a half years to this point and was in a beautiful relationship with Alex (the girl I initially ignored on the beach). I had even found time to fly to Cleveland to see her and meet her family.

She lived with her grandmother. I slept in her department-store-of-a-basement on a fold-out couch surrounded by purchases stacked to the ceiling. Her Grandmother made a hobby out of buying all sorts of things at a discount and gifting them to her family as needs or celebrations arose. It was impressive to be sure, but quite cluttered! In fact, there was a full bathroom in the basement I was unaware of because it was hidden by the stacks of purchases.

Grandma was a beautiful soul. Easy to talk to, very understanding of others, and protective, in a loving way, of her family. Her family are also protective of Alex. I understood that feeling and honestly liked the fact that they might be leery of me in favor of Alex. I had a fantastic visit and really enjoyed meeting her family.

Following "C" school graduation I was granted a week of leave (vacation) so I drove home. Alex drove to my hometown and met my family. In advance of her visit, I had told my family a "friend" was coming and would stay for a couple of days. I mentioned my friend's name was Alex but failed to mention my friend was a girl. My parents were fine with that, but were surprised to find Alex was a girl. To this day Alex reminds me of that. I thought it quite funny at the time and still to this day.

Perhaps a story is appropriate of Alex's visit to see me in that sturdy and cold abode we three men called our temporary home in the city of Virginia Beach. Obviously, it was winter. She couldn't believe how cold our apartment stayed, especially the kitchen. It just so happened that the kitchen sink developed a leak in the water supply line while she was there. I had an idea how to fix it (since calling the manager never got action). She and I went across the street and bought regular bubble gum. I gave her a couple pieces to chew as I did the same. Then I took our gum and used it to seal the leak. It worked!

She probably thought I was an idiot. I didn't ask because I didn't want to know the answer. Later that night we went to a horror movie. Returning to the apartment, I made her enter first. I told her that was because if something happened, I could run faster for help. I used that logic on several occasions. Again, she probably was sure I was an idiot. In fact, I suppose I am.

I joined the Navy to create a more positive direction and so far I believe that's happened. Up next was some time off at home. Then to Charleston, South Carolina and my submarine.

Opportunity or regret, like open water, lie before me.
Seize it or leave it, then maybe be free.

Chapter 8

Reporting for Duty

It's been one and a half years since taking the oath and enlisting in the Navy.

All of my prerequisite training is complete and I'm reporting to the Charleston, South Carolina Naval Base to start submarine duty. My assignment is on the USS Lewis & Clark (L&C), SSBN 644 Blue Crew. Upon arrival, I reported to what amounts to the human relations group. That's not what the Navy would call it, but that term is more relatable to civilians. My sub is in port, but I can't go on board. They'd been selected to launch some missiles as part of an inspection so they had to come back to port to trade out missiles to get the test missiles on board. Rules are that new personnel are not permitted to report on board until the launch and inspection is complete. So I'll be assigned to a work crew until that time comes.

During training at Dam Neck, Virginia, teachers emphasized and encouraged us to show eagerness to start and complete submarine qualification in one patrol cycle as the crew views that as very favorable. One patrol cycle was one patrol at sea and one shore period (it varies, but approximately two hundred and ten days combined).

A ballistic missile submarine operates differently than a fast attack submarine. Due to having two crews (blue and gold), I suppose, the roughly 105-day cycle was created. The blue crew would be assigned to the sub (or boat in Navy lingo) and then turn it over to the gold crew. I expressed my interest in getting on a sub to start sub qualification to the HR representative. I agreed to go to any boat, even if that assignment was temporary, or until the L&C was available. I was tired of training. I just wanted to do something more impactful. He was perhaps impressed because a lot started happening to get required clearances and checks done STAT. Other people's physician appointments were delayed to get me checked-out. I felt bad because I'd walk into a doctor's office and be taken immediately as others sat in the waiting area.

The dentist even commented, "Somebody must really want you." Everything went well. My last visit was with either a psychiatrist or psychologist, I was unsure of their title. However, I also didn't know or care regarding the difference. He and I talked for a while, but it was only two topics that are burned in my memory.

He asked if I had any moral objections to nuclear weapon use and asked me to draw a picture of a boy and a girl. I denied objections to the weapons, especially since my job on a boat was to launch those weapons if they were present and/or as directed. I always knew what the job might entail. I completed a quick stick figure drawing of a boy and a girl – that was kind of weird and uncomfortable, but he cleared me.

That HR person seemingly had a lot of pull. I've never had preferential treatment like this. These appointments being completed along with word that I had the necessary top secret clearance meant I could go at any time. Following the quick succession of appointments, I was placed on a work crew while awaiting word. I was excited and nervous not really knowing what to wish for. Would it be better to go to the boat I was assigned or do a temporary assignment on another boat? How would I be received in those two situations? I had no idea what to wish for. I thought it best to accept whatever is suggested or offered.

I checked into a barracks and was assigned to a room. The rooms had the same setup as my previous room at Dam Neck – a bed, free-standing locker and cabinet that opened to a desk. My initial impression of my roommate was that of a kind-of squirrelly dude. He was shorter than me and never stopped moving. Even when he sat down, something was moving – hands, arms and eyes were constantly in motion. He also talked a lot but we rarely had a real conversation. He'd just ramble on, but he was a nice guy as well. He was in the same holding pattern as me, but not waiting on a sub. His assignment will be a surface ship. Thank God. I got tired just being around him.

The next morning, I and many others mustered (or assembled) for our work assignments. Prior to this day and each morning afterwards I ironed my dungarees, adding military creases, and shined my shoes to be as presentable as possible. I didn't know what benefit it would have but thought it wouldn't hurt. Most did not use their time to iron and add creases to their work uniform. I mingled with the group while awaiting those in charge, but ensured I wasn't near my roommate. I couldn't take working with him all day.

I didn't initiate an introduction of myself to anyone, instead preferring to maintain my privacy, but some started a short conversation with me. Everyone seemed pretty easy to get along with. The work crew was separated into smaller groups and assignments given. All of the work was easy and I never broke much of a sweat. For the most part I cleaned bathrooms. I made them sparkle. I hoped someone would notice either my work ethic or pressed uniform. Maybe one of them will get me a better job or on a boat sooner. Thankfully, someone eventually did.

As was frequently the case, one morning I was assigned to clean a head (bathroom) in one of the buildings. While bent over cleaning one of the many toilets, the work crew team leader and another sailor surprised me, approaching from behind. He told me to leave everything and come with him. The sailor that accompanied him took over. I couldn't figure out what I did wrong. It turned out, nothing. He explained that he selected me to interview for a different position and led me to an office where a lieutenant wanted to talk to me.

Arriving at that office I met the lieutenant. He was seated behind a rather nice, though small, wooden desk. The team leader explained that he was impressed by my work ethic, even stating that I went as far as ironing military creases in my dungarees. Dungarees are a work uniform consisting of jeans and a lighter blue work shirt – under that we wore a white t-shirt. With that uniform we wore either the well-known round white hat or a ship's ball cap. The lieutenant asked some questions, ultimately explaining that he needed someone with a top secret clearance for two-man control. There was apparently a new rule that meant handling or viewing classified information required a second person to ensure that information wasn't compromised. This job was way better than the work party and scrubbing toilets except that I might be needed anytime, day or night. I didn't feel all that productive, but the job was easy – again, no sweating involved, but even this job only lasted about a week or so. My next assignment got me a little closer to my sub.

Luckily I like change because I'm being constantly reassigned. This time I'll report to the gold crew of the L&C. That's the other of the two crews for the submarine. Ballistic missile submarines are different from fast attack submarines in many ways, but one of the ways is in the way it's manned. As mentioned previously, a ballistic sub (like the L&C) has two crews: blue and gold. A fast attack has one crew. The two crews alternate being at sea and on shore.

I don't know the reason(s) for having two crews, but speculate it's to strive to have the sub at sea on patrol as much as possible. So while the crew I'll eventually join is at sea (blue), the gold crew was in port. For the most part I was a gopher. My job was making deliveries of paperwork or picking up mail, but I did learn the organization structure of the sub and a general explanation of the routine when the boat returns from sea along with the routine of being in port without the sub. The biggest benefit for me is it'll help kick-start sub qualification when I do join the blue crew.

Patrol One

It was late June when the L&C returned from its missile launches. They apparently did really well. I was glad of that because I assumed that meant the crew was good. The time in port was to be short, but it allowed me to get on board for the final portion of its patrol.

When the boat comes to port, it ties up to a type of repair ship for that type of sub. That ship is called a tender. So with my packed sea bag I walked the long (and wide) pier to the tender, crossed it with a guide and onto the sub's gangway to request permission to board and presented my paperwork. I never before felt so intensely nervous and excited at the same time. I felt I was stepping into the unknown even though school had prepared me well. The sentry called down to the FTBs to have me escorted. There seemed to me to be a lot of activity all around. The sentry kept asking me to slide one direction and then another to keep me out of the way. They were apparently scrambling to get back to sea. The top surface of the all black painted sub is not all that roomy, especially with the increased traffic. In fact, the sub itself doesn't appear all that big, but of course, in the water, the majority of the sub was below sea level.

Steve, one of the FTBs, arrived to guide me on board. He's a really nice guy – a tall, lanky country boy best characterizes him. His pants have extensions sewn to the bottom so them of his pant legs would reach his shoes. The slight grin he always had made me wonder what he last did to cause that. Luck was on my side as all the FTBs are a nice group. An excellent team that helps each other be as successful as possible.

I entered the sub through a hatch on the side of the sail with a diameter not much bigger than my shoulder width, meaning it was a struggle managing my sea bag. The sail is that part of the sub extending up from the deck. The fairwater planes extend out from it on each side and is where the periscope extends above. Upon descending the ladder and entering the sub, I immediately noticed the 'submarine' smell. It's not like a locker room - not objectionable - and I'd get used to it. It wasn't enough to cause hesitation, but certainly unique, noticeable, and memorable. By the end of patrol, all my clothes and hair retained that scent.

Steve and I snaked through very narrow passageways and down more ladders to the MCC (Missile Control Center). All the halls in a sub are only wide enough for one person to comfortably pass or two people turned sideways to traverse. Moving deck to deck required ladders or steep stairs. Most times the crew just lightly held the sides or handrails and slid down. Moving horizontally between compartments meant passing through smallish hatches requiring me to bend over to avoid hitting my head while at the same time stepping up and over the raised bottom of the door. I'm relatively short and eventually developed the ability to run/jump through those smallish hatches or doors when responding to emergency situations. Anyone over six feet tall had issues throughout the boat.

Entering MCC required knowing the secret code for the lock. It is a space with secured access since it is from there that the missiles are targeted and launched. Now in the MCC I met all the guys. Ken, a Mormon, was taller than me, heavy set, blond, easy going, easy to talk to, and welcoming. He has a genuine smile that puts anyone at ease. Denny was a short, stocky gentleman. He also was laid-back and very knowledgeable and smoked often. He doesn't initiate conversations, but willingly participates. Ted was tall, greater than six feet and heavy-set but, quite honestly, a kind-of space-head. His mind seemed to drift off sometimes or maybe he's just easily distracted. Nate was from Missouri and definitely a 'show me' state kind of guy. That's his description of himself when we met for the first time. He had a heavy accent, was energetic, outgoing, and a hard worker who used chewing tobacco occasionally.

After meeting the fellas, I was taken to the berthing area and assigned a rack (or bed). Storage was very limited. The three-inch-thick mattresses sat on a tray that lifted to reveal a three inch or so deep compartment that ran the length and width of the mattress. Above the mattress was a tray that could be lowered when not in the rack sleeping. It was perhaps two inches deep that extended to almost the length of the bunk. The rack and mattress itself was approximately three feet wide and six feet long and about shoulder width tall in space and a curtain available to pull for privacy. In reality, there is no such thing as privacy on a submarine.

I wasn't comfortable moving about the boat yet but did find my way back to MCC from the berthing area. I knocked on the door for entry since I hadn't been given the entry code. I was given several qualification (shortened form being qual) cards that I was to start (and finish as quickly as possible).

The 'qual' cards were for: MCC Technician, Nuclear Weapons Security Guard and Submarine Qualification. Things were moving quickly. It was easy to feel overwhelmed. I found myself taking deep breaths occasionally, but the FTBs helped me get organized and were full of great advice. I can't overstate how much help they were to me.

The date I reported to my sub and officially started was June 24, 1985. I had entered the Navy on January 3, 1984. In terms of priority, it was stressed I focus on MCC Technician first so that I can stand watch while underway and pull my weight within the department. Standing watch is comparable in the civilian world to working my assigned shift.

Then I was to focus on Nuclear Weapons Security Guard so as to be able to help when in port, onloading or offloading a missile, or whenever someone had to physically enter a missile either at sea or in port. Lastly was submarine qualification so as to be useful as a crew member. Reality was that I had to show progress on all three simultaneously.

In terms of something similar to a legacy, submarine qualification was by far the most important and meaningful of accomplishments in my Naval career. It was also a difficult and lengthy process. It meant understanding every part and process of the sub in detail. In retrospect, it also forces one to meet and talk with nearly the entire crew. The result is a sort-of brotherhood and respect that I held then and still do now. Broadly speaking, the qual card listed various aspects of submarines processes, equipment and administration that an orientee must study and then seek a qualified crew member to get that item 'checked-off'. I was told I was required to always have the qual card on my person. Any qualified crew member had the right to request to see it at any time and ask me questions. My process to get someone's signature was as follows:

1. read/study documentation
2. review appropriate flow charts and diagrams
3. walk and visualize each process to include equipment, piping, valves, ...
4. seek out a crew member for an initial review
5. re-study all available documentation and notes
6. ask a crew member to provide the check-off, answering any question he posed and looking up anything I couldn't immediately answer

Out of respect, whoever I asked for an initial review was who I returned to for the check-off. A crew member may ask anything, in any detail, prior to signing off.

I often had questions posed to me requiring additional study before getting the signature. I tried to seek out crew members known to be difficult. That way I was confident I knew the subject matter. An example of a more challenging check off to obtain was when I needed to be checked off on emergency equipment in the engine room. The crew member decided to test me by making me wear a gas mask (an EAB-something that I'll explain later) with the mask covered so I was blind.

I had to walk the engine room basically blindfolded and point out locations of all emergency lighting, fire hoses (and their length), fire extinguishers (along with extinguisher type), and other similar equipment. The idea is that I'd be able to locate the equipment even if a fire caused thick black smoke to fill the compartment. That was incredibly challenging, but I certainly had great familiarity with the engine room when he signed off. My goal for completing sub qualification was one patrol cycle. Since I started mid-patrol, I'd be excused for needing a portion of the next patrol at sea.

First-up for me was meeting 'Doc'. That's the lingo for the singular medical person on board. He was a kind-of EMT trained person. Not a nurse or doctor, but still quite capable. He was very welcoming, encouraging me to stop anytime to talk or ask questions. He also gave me the dosimeter I was required to wear anytime on board the sub. At the end of each patrol it was checked to determine radiation exposure. The submarine was nuclear powered and had the capability to carry/launch nuclear weapons, but he let me know my radiation exposure would probably be less than what I'd get if I spent a day at the beach in the sun. I was starting to feel overwhelmed meeting everyone. I'm terrible at remembering names. Thankfully, uniforms have last names on them.

I kept repeating in my mind that I'm not the first at this experience. If others could do it, so can I.

The day's fast pace continued in the chow hall (where we ate). In terms of lingo, the galley is the kitchen. Entering the chow hall for the buffet-style meal meant standing in line that often extended into the tight passageway. Sometimes it even extended out of the chow hall, down the passageway and down the ladder by MCC. We'd be given servings of food we requested, get our own drink, and then sit at tables with bench seating, but it was not presently mealtime. I was brought there to listen to the COB (Chief of the Boat). He was a somewhat portly man, around five feet eight inches tall and very approachable and easy to talk to. He is the top enlisted person on board. Among various topics, he described the upcoming patrol in very broad terms.

Within his presentation was a statement to see him if we'd like to volunteer to work topside when the boat departs and when it returns from sea. This group handled lines to tie-up the boat when returning from sea or releasing the lines for departure. Following the COB's presentation, I followed him the short distance to his small office to volunteer. He wrote down my name and told me I'd be notified when to report to the chow hall for instructions.

Finding my way back to the MCC, I entered the code to unlock the door and entered. My fellow FTBs greeted me with smiles. They were truly a great group. I find myself expressing that thought repeatedly, but I can't relax yet. My day isn't over. I just learned I have to meet with the weapons officer or WEAPS as everyone calls him. I had to be shown the way to the stateroom he shared with another officer. As everyone seems to be, he is easy-going. When he's joking around, he gets a kind-of shit eating grin. He always made me smile with his dry humor. This meeting was a discussion of patrol in a very general way, but he did ask me if I had any moral objection to nuclear weapons. He became the second person to ask me this and I again denied. Thankfully I didn't have to draw people again. It was a short meeting. It sort-of seemed to me to be something that was necessary so it could be checked off some list, but I was glad to formally meet him.

The remaining time of this first day, along with the next

two days involved a tour of the boat, helping anyway I could and starting my qualifications requirements. Nate is going to be fun I think. He always has a smile and a joke, along with an incredible work ethic. He often removed his dungaree shirt in port when working, instead opting for a white t-shirt with cut-off sleeves since he apparently sweats so much.

The day arrived when topside volunteers are to meet in the chow hall at 1400 (2pm). I'm a couple days into submarine service and starting to feel comfortable moving around, albeit to a very limited degree. Everyone in the room is sub qualified. I'm the only non-qual, but I'm welcomed to this crew.

We watched a short video of safety precautions related to line handling. The COB made an entrance to emphasize safety. Doc also made an appearance in that respect. I met the other men in this group and was glad to receive comments from them to just stay close, they'd ensure we'd be safe. We were shown how to make a few knots, but I honestly quickly forgot. Everything is happening so fast. The short time in port has quickly come to an end.

Finally, I would soon be going to sea. Tomorrow, the lines (ropes) keeping us tied-up to the tender will be released and tugboats will direct us out of the bay to the river where they would leave us. We'd transit down the river under our own power, leaving the area of the Naval Weapons Station and pass many backyards of city residents. Some would be out and waving. As we move down the river the topside crew will stand and occasionally wave. We will be wearing life jackets and be tied-off to the deck to prevent any possibility of falling overboard.

Eventually reaching the mouth of the river we'd be in view of the beautiful and colorful downtown of the city of Charleston. There is always a smattering of small watercraft. The people on board them, I'm told, always seem surprised to see us. They stop whatever they're doing to wave. This routine would be reversed when returning to port. I'm soon to be topside for all this to play out. Everything related to 'shoving off' occurred as described. I loved waving to the residents and boaters of Charleston. My fellow topside crewmembers performed perfectly. I was glad to have them by my side.

We are now passing through the mouth of the river with just the open sea ahead. In the back of my head were feelings of nervousness. The time has come to perform. I hope I can do it. Very soon, I will finally be at sea. The time has come to secure topside. The rope lockers holding the lines used to tie-up in port are sealed tightly. A second person verified it. This is required to ensure no rattle will be present when submerged. Silent operation when submerged is a top priority. Then, in single file, we re-entered the sub. The last person was the officer in charge topside. He ensured no personnel were left topside, then shut and tightly secured the hatch. Life on board is a whole new world.

The adventure begins. I've heard the saying, be careful what you wish, it might come true. No use looking back. I experiment with serious looks to hide my nerves. I will focus on grasping every opportunity to learn.

Passing optimistically from light,
Ignorance deceives one into thinking bravery

Chapter 9
Finally Underway

Leaving topside, I descended the ladder to operations upper level, taking the ladder step by step rather than sliding down like I observed the crew doing. I can now confidently get myself to certain areas of the boat – the three levels of the operations compartment are some of those areas and also my main stomping grounds. After stowing my life jacket, I made my way to MCC in Operations Lower Level, I sat with the other FTBs and listened. The MCC is one of those areas with access to all the various communication channels on the boat, thus allowing us to listen in as necessary or desired. With all personnel safely below deck, the captain gave orders directing the boat to open waters.

We would dive as soon as we could to avoid any possible Soviet "fishing" boats. They always sat offshore. The 1MC is the main overhead announcing system. From it came the order, "Dive, Dive, Dive". We would soon be at depth and cruising comfortably. Being submerged made everyone more at ease. A submarine on the surface is just not natural. While we were transiting across the surface I moved about the boat a little to witness the lifelessness.

While we're on the surface, everyone just sat and waited- not much talking. And no real interaction except that needed for official business. Being submerged brings a sort of comfort. That's where a submarine is supposed to be. That is its environment and where everyone feels safest and being underwater allowed everyone to relax a little. After settling into our cruising depth, Denny, sitting in the MCC with his feet elevated suddenly arose saying, "It's that time fellas."

It was time to change into poopie suits and develop a routine. How they got the name poopie suit is beyond me. I feel odd every time I say it. A poopie suit is a long sleeve Navy blue one-piece uniform with a long zipper in front running from the crotch to the collar. All appropriate Naval insignia was sewn on. It and tennis shoes were the daily uniform at sea. The tennis shoes were to reduce sound associated with moving about the submarine. Everything necessary and possible was done to reduce noise.

The poopie suit was comfortable and easy to get on quickly when leaping from one's rack during battle stations or any emergency situation. The dungaree working uniform wouldn't be needed again until we returned home. The poopie suit was never worn where the public might see it. Soon all of us not on watch left for the berthing area to change. As I previously mentioned, standing watch is akin to civilians working their shift. When underway, we did six hour watches. The other significant adjustments were to switch to Greenwich Mean Time (GMT) and to an eighteen-hour day. Therefore, we'd be on watch six hours, then fit everything else into the remaining twelve hours. That means sleep, shower, and additional work related to collateral duties and drills.

Being a non-qual, I rarely slept more than two to three hours straight. Instead, I got catnaps when I could and drank a lot of coffee to stay awake. The submarine is where I was introduced to and got addicted to coffee.

GMT meant we switched to be the same as Greenwich, England. This allowed all Naval craft to be on the same time. Several times during this first cruise of mine I found myself wondering and asking if it was day or night at home and what day of the week it was. It was easy to lose track of the days when not having access to the sun.

I returned to MCC wearing my new poopie suit. “Well, I'd say it's time you got started,” said Steve. He asked for my MCC technician qual card and started peppering me with questions. Honestly, this qualification wasn't particularly challenging, especially compared to submarine qualification. After all, I had recently just finished school so the information was fresh. We walked around MCC. MCC is a secure, air-conditioned and carpeted space with two comfortably padded captain's chairs that could rotate 360 degrees and were positioned at the control console. Working here was much more comfortable than any other space on the sub. The several aisles of computer hardware contained the missile launch system computer. Steve and I strolled the aisles as I identified and explained the purpose of each computer subsystem behind each door. We ended up at the control panel from where any launch or maintenance test was initiated and carried-out. Each button on the panel was labelled.

“One of these days,” Steve said, “you're going to know every button from memory.” That test of memory would come much later when qualifying as MCC supervisor.

Steve and I talked a while. He signed off on a number of items on my MCC technician qual card, but eventually tired or got bored and wanted a break. That's pretty much how several of my initial qualifications went. I'd get as much signed off as possible from whoever I could until they wanted a break. Each day of my first patrol would find me studying, then seeking an appropriate person to sign-off that topic.

Between MCC technician, Nuclear Weapons Security Guard, and general submarine qualification I move between each as qualified personnel were available and willing. Poopie suit pockets were luckily large enough to carry several multi-page qual cards. Life aboard a submarine required considerable adjustments.

Once I left topside, I knew I'd probably not see the sun or breathe fresh air for perhaps multiple months. I had no idea how many days were left in this patrol so was unsure how long we would be at sea. I had never experienced darkness like when lights are off in a submarine. With lights off, my hand only inches in front of my eyes could not be seen. But in spite of that, as time progressed and I became more and more qualified, I acquired the ability to move in utter darkness without difficulty and at speed. Meals on board were good; although fresh milk and salad were exhausted in a week.

Once per patrol we'd have steak and another time crab legs. I've heard food on a sub is better than surface ships. I believe it but cannot say with certainty having never served time on a surface combatant ship. On a sub, some sort of meal was served every six hours. Breakfast was at 0600 (6am), lunch at 1200 (noon), dinner at 1800 (6pm) and midrats at 2400 (midnight).

Midrats generally consisted of reheated leftovers. Lights in berthing (where not everyone, but most slept) always remained off except for the once weekly tradition when everyone was awake to clean assigned areas. The oxygen level in the boat would sometimes be slightly increased during those times so everyone would get a little boost of energy for cleaning.

The relationship between enlisted and officers changed in a positive way once at sea. At sea, the traditional discipline associated with the military changed to a more relaxed posture, yet also associated with more direct personnel interactions. If someone messed up, they were told face to face without regard for rank in a professional yet frank manner. Respect was maintained, but little effort was wasted considering someone's feelings.

Water use was watched very closely. After all, we had to "make" our water so conservation was paramount – for God's sake, don't waste it. Our submarine shower allowed for only fifteen seconds of running water. In fact, for a couple days during every patrol, someone would be assigned to the head (bathroom) with a stopwatch and chair to time individual showers. Anyone using more water than allotted were reported.

Somehow I grew use to cold showers. Mine consisted of an initial three seconds to get wet, then soap up and twelve seconds to rinse. The nuclear reactor always had priority with the available fresh water so every patrol seemed to have one week with no showers permitted so there would be plenty of water for the reactor and for cleaning.

When not curled up somewhere with a pot of coffee studying, I sought the solitude of MCC. It gave me a chance to relax a little since it was somewhat more familiar and relatively private. Jason, one of the FTBs I hadn't mentioned yet, was always a good source for advice. This was his last patrol. I suppose I would be his replacement. That thought added pressure. He'd be discharged soon after this patrol. He had done enough patrols that they had become routine.

During the drills when we practiced missile launches (referred to as Battle stations, Missile), I was ITOP. ITOP was an acronym for Integrated Test Operating Panel. From it, we ran preventive maintenance routines. It was also where I stood during battle stations, immediately to the left of WEAPS as he sat in one of the captain's chairs. To WEAPS immediate right, seated in the other captain's chair, was the launch supervisor (enlisted). ITOP would be my job during a missile launch to coordinate communications between the missile compartment and MCC. I was sometimes the person who communicated orders from WEAPS to the missile compartment and reported any significant activities occurring in the missile compartment to WEAPS.

Jason was always good for a friendly and informal critique of my performance. For which, I was appreciative. He had a confident air about himself that some I think mistook for ego.

WEAPS also was helpful. When he spoke, it was in a low volume and calm tone with a sort-of raspiness in his voice. In fact, he sometimes neared a mumble. I wore my headphones with only my left ear covered and my right ear open so I could hear WEAPS when he spoke. He'd occasionally lean into me with a casual recommendation. All of which I put into practice.

We got to the point where we'd greet each other in passing with a "What's up?" Calling him 'Sir' wasn't natural. Some in the military find that style of communication between an officer and enlisted unusual, but then the more casual relationship between officers and enlisted on a sub, especially when at sea, is different than on a surface ship. And then there was Denny, who I stood watch with. No matter the circumstance, he was never in a rush, but could troubleshoot anything in MCC. I qualified as MCC technician during this abbreviated patrol because of him. He wasn't the type to go out of his way, but willingly did anything asked of him. All of these qualifications I was seeking got me out and about in the boat.

Submarine life is not just close quarters physically, but inevitably also results in intimate knowledge of each other's lives. Nothing was off limits nor were blunt questions and answers. The face-to-face interactions on a sub were far different than in civilian life. We had no boundaries and rank had little privilege. In this first patrol I learned of some pranks.

However, the knowledge also required unspoken devotion to the secrecy of those committing the pranks. Being new, I wasn't comfortable participating, but would end up doing so in later patrols. The periscope, a tool of the officer in charge, gave that person a sense of importance and privilege. After all, in normal circumstances, only he could peer through it. Some let that privilege go to their head. The officer standing watch as officer of the deck was sometimes fairly new and as such suffered the consequences from seasoned enlisted personnel having fun. In this case, unfortunately, there was no way he could know someone had "polished" the periscope with black shoe polish. After using the scope to view the ocean's surface and/or look for contacts (ships or aircraft), he'd be stained and the butt of jokes for his raccoon eyes.

We also performed indoctrination rights in some new officer's staterooms. The recipient noticed these efforts after his six-hour watch when he'd retire to his bunk only to lay his head on a frozen pillow. Someone had doused it in water and froze it. The life of an officer had its perils, but what gave me the most reward was stealing Fritos and caramels from under the noses of the chief petty officers.

They loved watching us cleaning their quarters as they sat. Why they didn't realize that stealing the snacks in storage was why we eagerly anticipated cleaning their quarters I couldn't figure out. They obviously had lost their edge. Submarine life meant entertaining ourselves and nothing was more entertaining than greasing the handrails of the steps leading from Operations Middle Level to Operations Lower Level.

These steps experienced high traffic as personnel left the chow hall after a meal and headed down to the berthing area. No one walked down steps on a sub. We would merely lightly grasp the handrails, lift our feet, and slide to the bottom, but when those same handrails were greased, men sped to the bottom with such speed, force, and lack of control that they crashed into the wall below. We laughed uncontrollably. Thankfully, no one had long term injuries.

We've been on patrol for many, many weeks now and I haven't had four straight hours of sleep since the start of it, but it'll be over soon! The constant need to get some part of qualifications completed has kept me busy and I finally completed one. I'm MCC technician qualified and can't proceed any further with Nuclear Weapons Security Guard until I qualify at the firing range.

Being a qualified technician makes me feel part of the crew in a small way. At least now I'm productive with Denny and I together on watch-he usually with a cigarette, relaxed, and his feet up. I perform the scheduled computer maintenance and answer any calls. Most times, our watch is a bit boring. The only excitement comes with battle stations-missile drills. All other drills when on watch, including fire and flooding aren't very demanding unless the drill leaders decide the catastrophe is in the space we're responsible for.

I continue my sleepless routine with submarine qualification. Tim, a torpedo man, let me use a small spot behind some equipment in the torpedo room as a private place to study. One of the cooks looks the other way, so to speak, as I make and take a pot of coffee with me to the torpedo room. I made myself comfortable on the deck surrounded by books to study and the pot of coffee with a coffee cup.

Crew members entered and exited the torpedo room never aware of my presence. Tim never told anyone of my routine. This gave me the privacy that's so hard to find on a sub to study. I was a zombie way before they were popularized.

I responded to battle stations, flooding, fire or other drills from that spot. I got more than a few odd stares, as an FTB, responding from the torpedo room. In fact, I received a job-well-done from one drill leader due to responding so quickly to a fire drill in the torpedo room. I chuckled slightly as they remarked, "...he came from nowhere" in responding so quickly.

My submarine qual card is three pages long. By the end of this patrol I completed one and a half to two pages. That's pretty good progress. Waiting for the next time we're at sea to complete it isn't ideal, but unavoidable. The COB scheduled a meeting for topside volunteers. The boat will be approaching Charleston in two days so it's time for a review of procedures, especially safety. I gave myself those last two days off from qualifying efforts. My fellow FTBs tried to set my expectations for when we arrive in port. There'd be a group of wives and girlfriends, some bringing kids.

I was one of the few with no one to welcome me home but standing to the side watching the reunions made me smile. The last day was finally upon us. I waited in MCC for the time to muster (meaning to gather) and go topside. This time I was in my dungarees, rather than a poopie suit. I had been sitting and waiting for a couple hours watching the clock. Anticipation of our return wouldn't allow for sleep.

At the directed time, the topside crew met in the chow hall. All in dungarees and a ball cap with a submarine insignia and the boat's name on it. We each put on a safety vest with a lanyard and a life jacket. The lanyard provided the means to attach topside and prevent an overboard incident. As a group we walked from the Operations Middle Level Compartment where the chow hall was to Operations Upper Level and the Conn. It's at the Conn where the OOD (Officer of the Deck) operates along with those driving the boat and performing navigation, radio and sonar duties in nearby rooms. From there we climbed the ladder to exit through the hatch to topside.

I immediately noticed the smell of fresh air and the blinding sun. I didn't expect fresh air to have an odor. It has been so many weeks since experiencing either. Both were beautiful! I didn't have sunglasses, only the Navy-issued round wire-frame glasses. They were quite thin and flexible so as to bend and allow for a good seal with a face mask.

As when we left for sea, people on boats and land waved, this time to welcome us back home. In port, we tied up to the tender (submarine repair ship) rather than directly to the pier. Over the next several days we reported off to our counterparts on the Gold crew. Everyone on the sub had their mirror on the Gold crew. This transition was easy and limited to several eight-hour days. Once the Gold crew took over, we started several weeks of R&R (rest and relaxation). The R&R routine required reporting in person once weekly. Other than that, we were off without the need to take vacation time.

As a storm's effects are revealed by a new sun,
The unknown will soon come.

Chapter 10

Reappreciating Land

R&R while living in a barracks has downsides. I had to deal with a roommate that was not of my choosing along with responsibilities of cleaning a room and barracks to Naval standards. That meant an abnormal attention to ALL horizontal surfaces. That could be a bed frame, curtain rod or even the top edge of the door and it's molding, but luckily, my roommate wasn't all that bad. Hopefully he felt the same of me. Although, he had this room to himself for quite a while until I got there so I would understand if he didn't welcome me. Honestly, we didn't see that much of each other. He had regular responsibilities on base since his Navy job was shore-based and I tried hard to get away from base daily, but I was required to commit a couple of hours a week helping clean the common areas of the barracks. It was easy work but living off base was more desirable.

Still, I was able to become a regular at a couple beaches. At least once a week I called Alex, which required a short walk to a pay phone. She and I maintained a long distance relationship since meeting in Virginia Beach. In some sense, these calls were our version of dates.

She's at home in Cleveland and working at a drug store but doesn't see this as her long-term future. I don't either. She continues to live with her grandmother who is quite nice, but as with me, Alex wants her own place. To call her I have to walk to a nearby pay phone that's maybe half a block away. It sits under a streetlight that is lit 24/7 and where mosquitoes seem to swarm me during my calls. They pester me whether it's day or night. How odd I must look talking while swinging my free arm and sort-of hopping to keep my legs moving. The damn mosquitoes won the battle each time though. I'd jog back to the barracks after each call desperate to retreat from their onslaught. My calls to her were all collect so they were short and sweet.

Back in Virginia Beach we somehow knew we'd marry. We decided it then and haven't changed our minds. What she didn't know is that I had purchased the ring already.

The first three weeks of the 105-day period (approximately) off the boat moved quickly. It was followed with weeks of classroom training delivering further detail into particular subsystems of the missile launch system computer and giving me more insight into programming. Additionally, I learned to properly conduct emergency repairs to the computer system using solder and wire-wrap techniques. The classroom time was spaced with a couple of days on the firing range and a week of hands-on practice in the simulator launching missiles. The firing range consisted of shooting a .45 caliber handgun and a twelve-gauge shotgun. Those were the weapons on board the boat. The handgun requirements I suppose I met having hit 21/25 shots on the target. I wasn't a sharpshooter but didn't need to be. I've had minimal submarine experience but couldn't imagine a circumstance where I'd be required to shoot for distance given the close quarters on board. The shotgun we shot strictly to experience it. We shot into a hillside. So I can say I'm adept at shooting dirt. For both weapons we also had to demonstrate loading and use of the firearm safety.

On this particular day I shot the shotgun next to a woman who worked administration on the tender. She slyly whispered to me, asking where the safety was located on the shotgun. She then copied my actions as I loaded the weapon. I checked the safety on my gun in full view and in an obvious way so she could see. We both passed. Then there was the week in the lab practicing battle stations-missile on a full-scale MCC simulator. WEAPS and all the FTBs were required to gather for this every on-shore period. We repeatedly ran battle stations-missile for eight hours, every day for a week. The instructor "broke" something different each time to create varying scenarios. Every scenario was followed with a debrief to review our strengths and weaknesses. No one was spared from critique. Officer and enlisted equally shared in congratulations and criticisms. The only requirement is that all discussions stayed in the room – privacy was strictly enforced.

At the end of the week we, as a team, had to pass a very involved scenario involving multiple troubleshooting challenges and scenarios in front of several instructors grading the group of us on a pass/fail scale. Thankfully we never experienced failure during my time on the boat. I loved this training. It greatly increased my self-confidence and my faith in my shipmates. A fantastic team-building experience.

Alex and I continued to talk each week. I learned to wear long pants and shirts during the bloodbath that was my time at the mosquito-infested telephone. I would be going to sea again soon. I didn't and couldn't say when to her as that information was confidential, but upon my return it would be a year since our initial meeting. We decided she'd move down during the R&R period following this next patrol, but planning it was impossible because I couldn't reveal when I'd be returning. I just figured we'd work it out later. I was honestly nervous and excited about this commitment.

Of course we'd marry. Going to sea meant being completely out of touch with her. The last time I was out of touch was with Ellen. I hope and pray things go well for me at sea and her at home, but also for both of our families. As before, I have no control. That breeds tension I didn't need. The Gold crew would be returning in days. FTBs, along with every other group, met to organize the turnover of the sub to us. We detailed the process to ensure we would get what we needed in the most efficient way before the Gold crew left. On one day before they returned, the entire blue crew sat in a small auditorium as the COB, captain, and others outlined the upcoming patrol. It was all very general.

FTBs were reminded to be aware of our travel habits, like ensuring we varied the routes of our travels to and from base in case someone was watching. The comfort of normal day-to-day living is starting to seep away. Sleep is somewhat more restless thinking about forcing myself to initiate an unpredictable sleep pattern on board the boat again.

Patrol Two

The boat tied-up on the outboard side of the tender, as is customary, and we, the blue crew, came aboard to welcome the gold crew's return and begin turnover. This, of course, was after they reunited with family. They provided a status of the weapon computer system, MCC itself, and all our collateral duties. Work orders were prepared as needed to get the services of the repair ship (tender). In terms of the broader picture, the next month's activities were to perform many preventive maintenance computer routines, ensure all corrective maintenance is accomplished, help load food stores onto the sub, and prepare for sea trials and patrol. The work schedule meant port and starboard duty stations. That means working around the clock every other day. One day, half of the crew works 7am-5pm, then went home. The rest worked through the evening, night and the next day. In that next day, the half that went home took their turn working from 7am until the next day at 5pm – meaning working all day, evening and night and the next day.

That alternating work schedule continued for about a month or so until time to go to sea. I was qualified as MCC technician and now finally as Nuclear Weapons Security Guard. That meant being busy with work necessary to be prepared for patrol. Some days were dominated by the process of onloading or offloading missiles. That required tight security involving many Navy and Marine personnel, but there was one particular incident involving a kid (by my eyes). My guess had him at perhaps seventeen, or maybe slightly younger. He was driving a small motorboat into the bay where our boat and the tender docked. He was very slowly creeping toward the sub with a girl onboard seated next to him.

Both in bathing suits slowly approaching, probably curious as to the happenings on the sub. I didn't perceive a great threat, but they were getting a little too close for comfort. We were engaged in moving a missile. At their closest, they were probably thirty yards or so from the tip of my shotgun – that's my guess to the extent memory serves. My nervousness was increasing as they creeped closer. Their speed was just slightly greater than drifting – quite slow. Marines were also providing security, and unknown to the small boat, were everywhere. Many in places unseen, but also stationed to be very, very visible. All were aware of the small boat and were on edge. Some Marines, also in a small boat, but with a machine gun mounted, ordered them to halt three times using a loudspeaker well before the kid got uncomfortably close in my mind. It seemed the kid was completely clueless as to the danger he put himself and the girl into.

I couldn't imagine them ignoring or not being aware of those Marines. Thankfully they listened the third time to the Marines since the kids' ignorance made them seemingly unaware of the number of loaded weapons pointed their way, mine being one of them. I had kept the safety on the weapon engaged, afraid of an accidental, though unlikely I supposed, discharge of the shotgun. Very slowly they turned and crept away. I can't count the number of prayers I privately said asking they stop, turn around and drive away. The stress associated with the thought of being possibly ordered to pull the trigger on the unthinkable was intense.

The exhausting month of alternating port and starboard duty was coming to an end as sea trials approached. One day was scheduled for the entire crew to help load stores (food and supplies). The FTBs traditionally found a way to assist inside the boat to avoid the sweat associated with the Charleston heat.

I volunteered to help the machinist group topside. My intention was to establish a rapport and friendship to ultimately help with submarine qualification completion when underway at sea. They were the best resource regarding many submarine systems. The only thing I really got was a severe headache absorbing the sun's intensity nearly the entire day.

Beginning nearly a week's worth of sea trials meant patrol was pending. The days allotted to sea trials were reserved to test every aspect of submarine performance to ensure we were ready to complete a successful patrol. It lasted approximately a week – but could vary significantly. For me, angles and dangles was what I looked forward to. It was a test involving diving and coming back up at steep angles. Anything not properly secured would be shook loose. As an aside, it also revealed anyone susceptible to getting seasick. We laughed at the angles we could lean back and forth without falling. Other significant requirements was taking the boat to various defined depths.

Most people are aware of periscope depth. However, there are others that are just as important. Of major significance is descending relatively close to crush depth. The name perfectly describes it. We didn't go to it, just close. It was a serious matter that demanded and received serious attention. There was no idle chatter among crew members, whether they be on or off watch. When back in port from sea trials, anything that broke or needing attention was quickly addressed in order to set out for patrol as quickly as possible.

I mailed the allotted six family gram forms to Alex and ensured I had everything necessary on board. Each family gram could contain at most forty words, with punctuation counting. The person sending the family gram must send it to the on-shore crew (Gold crew for me since I was at sea and part of the Blue crew) for review. It was reviewed by at least three people before being approved to be sent to the sailor at sea. The meticulous review was to ensure the person at sea doesn't receive upsetting information without leadership being aware.

At sea, no return messages were possible.

The next time I step on board the submarine, I won't step off for something like two and a half months. During that time it would be as if we didn't exist-silent in every regard. I and my fellow FTBs went out the last evening before shipping-out for one quick drink, knowing it would be our last for quite some time. We didn't stay long at the bar. Everyone understandably wanted to be home early enough to spend time with family and spouses.

At one establishment we sat outside and enjoyed the small acoustic band playing that evening. We selected seats on the bar's enormous outdoor deck at a large round wooden picnic-style table with an equally big umbrella as it was able to accommodate all six of us there that evening. The conversation was casual. The upcoming patrol wasn't one of the subjects. Steve casually pointed to a button on the umbrella pole wondering as to its purpose. To find out, I pushed it. The result was to cause the umbrella to collapse around us too quickly for anyone to react. We were suddenly all viewing each other in the darkness from under the umbrella. People around us laughed-but we didn't. Struggling, we were eventually able to raise it back open.

I'm sure the band didn't appreciate the distraction we produced. Not long after that embarrassment, we finished our drinks and left. I hope this is not a harbinger of events to come. With Alex in my thoughts, I now appreciate a little more what others feel when leaving. Going on patrol is putting aside everything and everyone to do something few understand or think about.

Does this worsen with each patrol or does it become routine? I'll eventually know the answer.

Never dream without others in mind.
Be the spark others look forward to find.

Chapter 11

Realizing the Goal

I volunteered to work topside again as we were soon to head back to sea. Leaving this time was a little more emotional. It meant something to me this time to not go below deck until the last possible moment. The moment played itself out as previously. We waved to the public and then eventually secured topside to start patrol. For those civilians witnessing our departure, waving was a story to the next person they saw and then we were forgotten. That's not to blame them, just to state reality.

My highest priority now is to finish submarine qualification. Modifying my routine now means standing watch for six hours, participate in whatever the daily drills would be, then study as much as possible in the remaining eight hours or so of our eighteen-hour day. Drills on my boat were done daily with few exceptions and lasted about three to four hours. Normally drills were helpful in ensuring everyone knew what to do in an emergency, but they also helped pass the time. Being underwater continuously for perhaps two and a half months (give or take) and essentially cutoff from humanity would become problematic if we didn't keep our minds and bodies busy.

I got permission from Tim to use some private space in the torpedo room again.

The torpedo men were very accommodating. The cost to me was promising him he would check me off on emergency equipment located in the torpedo room and operations compartment. That entailed reciting from memory to him the location and type of every fire extinguisher, location and length of each fire hose, emergency lighting locations, air damper positioning to evacuate the air in the compartments and EAB connection locations – just like I had to do in the engine room. All this was just a starting point to get the checkoff. This checkoff was time consuming, but in the end I was appreciative of his demands.

Perhaps a description of an EAB is now appropriate. They are an Emergency Air Breather. They consist of an airtight mask and a hose. The furthest end of the air hose had a quick disconnect fitting to connect to the many locations on the air manifold that runs through just about every space in the boat. Therefore, one would put the face mask on and connect to the air manifold using the quick disconnect fitting in order to breathe fresh air. Some EABs had longer hoses and personnel could daisy-chain off each other to form longer lines. To move about the sub, one would take a deep breath, disconnect and run to another connection. Memorizing connection locations was very beneficial. The relatively short hose made it difficult to do much. Obviously, if one was the first person in a line of people, it was protocol to ensure everyone in line was aware before disconnecting.

My vision problems required that I wear submarine glasses when underway. Actually, I wore them all the time. They were round, flexible metal frame glasses and easily bent to allow one to get a good seal with a facemask. This type of intense checkoff wasn't really unusual. We had to know everything in detail to accomplish tasks in the dark. So for those remaining eight hours or so daily, I prioritized my activities as follows: submarine qualification, collateral duties, eat and lastly sleep, but my plan was altered by the Captain in an unexpected, but fortuitous way.

I and several other non-quals (not submarine qualified) were put on a four section sparkle team. Sparkle team meant we were now a cleaning crew – "... make things sparkle". The four section duty meant we were the only people on the boat that operated on a twenty-four-hour day. We still had to participate in drills but were temporarily excused from our normal duties so as to clean for a six-hour shift. Then we'd have eighteen hours for qualification, collateral duties, eating, and sleeping. I didn't sleep more. Rather, sub qualification was accelerated. Many were jealous of our four section duty, but none wished to be part of the sparkle team. All of our cleaning was in the engine room and other machinery spaces. Each six-hour shift left us incredibly greasy, sweaty and filthy.

The purpose of our unusual group was to prepare for an inspection expected to happen near the end of patrol. The inspection was to focus on only the engineering spaces, sparing those of us on the weapons side. I cleaned with earphones in place, listening to music and thinking a lot about Alex. I nervously looked forward to her moving down with me. The nervousness, in retrospect, was due to my expectation that something would go wrong. It sure did with Ellen and in a very different way with the girl who so easily disposed of my gift of roses.

The four section sparkle team lasted about a month. I then returned to my normal duty and eighteen hour days, but by this time I had all the check offs for my sub qual card done and ready to be scheduled to sit before the board for a final comprehensive examination. For me, the board consisted of an officer, chief petty officer and two additional enlisted personnel. The format had the four of them on the opposite side of a table from me. I believe this examination might be slightly different depending on the submarine. I was to bring a blank pad of paper and a pen to make any drawings they requested and to document look-ups. They could ask anything and in any detail about the workings of the submarine. In my memory I believe it lasted approximately an hour to hour and a half, but my fellow FTBs prepared me with some great advice that I used then and in every interview throughout my life afterwards.

They suggested:

1. Answer only what's asked. Don't elaborate.
2. Take your time.
3. Ensure you understand the question before answering.
4. Ask the panel to elaborate to clarify questions, as needed.

I was incredibly nervous. They expected it. After all, each of them had been in this position at some point. I answered every question and addressed every scenario poised to me, but in the end had two look-ups – meaning I'd get my dolphins only after I got back to the appointed panel member with answers to those look-ups. It turned out, I initially provided the correct answer during my examination for one of the look-ups. It was the chief petty officer on the board that was wrong. That left one look-up which I easily provided. The Captain attached my dolphins insignia pin to my uniform during the pinning ceremony and I was almost overwhelmed with relief, pride and satisfaction.

Of course, shipmates quickly brought me back to reality when they "pinned" my dolphins by forcibly punching my metal dolphins. Pain and a bruised chest was the result. This pinning tradition has been in place for many years I believe. That lasted for just a few hours. Each shipmate I encountered had that opportunity if he chose it – not all did.

I met my goal in the timeframe I planned. I thought I could

relax the rest of patrol but was mistaken. I had been slow-walking my qualification as MCC supervisor and my fellow FTBs were patient, but that now had to change. Progress had to be made quickly now since Denny won't be back after this patrol. Qualification as MCC supervisor came after a few weeks of intense work. Luckily, I didn't completely disregard it previously so I had only a small amount to complete. Becoming supervisor meant not just being technically sound as it related to the launch computer system and missile guidance system, but also thoroughly knowledgeable of many protocols and policies related to MCC and the missile compartment.

Some of the reference material was classified, but the job required such intense familiarity with the content that I had some of it memorized. I finished patrol standing watch with Denny, but now we both are supervisors. He out-ranked me but allowed me to assume the supervisor role to gain confidence and experience. I don't think he ever got enough credit for his leadership.

Somewhere in the middle of finishing my MCC supervisor qualification, halfway night came and went. As the night implies, it marks the halfway point of patrol. It was noted in very low key style by the crew – at least that was true on my boat. A few sat in the chow hall together and sang a little as someone played guitar. Others relaxed and read books they had brought and still others re-read the family grams they had received to date. No drills were scheduled. Everyone relaxed for this one day.

I personally always had a rough time with the halfway point of patrol. All I could think about was the people I was missing and how many weeks still left on patrol. I never found a good way to handle it. I found it very difficult to relax, even for this one day. We could receive a maximum of six family grams at sea. We all looked forward to receiving them. They are really our only connection to loved ones and the outside world. Especially since we are unable to send similar messages.

I gave all six to Alex and she thankfully sent all six to me in every patrol. In one instance, a family gram sent to me was incorrectly interpreted by everyone to think Alex and I just had a baby. The Captain, with his entourage, hand delivered the family gram to me and congratulated me. They were surprised that I never passed along that she was pregnant. I read the family gram and corrected their interpretation.

In actuality, it was a friend that had the baby. It turned out to be a comical error.

This patrol passed without an incident with me as supervisor. My duty routine changed in this new position. I now started my six-hour watch with a visit to navigation to understand where the boat was located geographically and where we were headed, but my watch was generally quiet. I was usually able to relax in MCC. There was the occasional radio message calling us to a battle stations drill as well as our boat's self-initiated drills, but Denny and I spent the majority of our time talking about family and life in general. Days passed more slowly now as the patrol was easing toward its end. I tried to avoid this trap, but inevitably I and everyone increasingly started thinking and talking about getting back home.

The news was passed that we will rendezvous with a small surface boat to allow an inspection team on board. The inspectors were to conduct the expected inspection of our engineering spaces. We all knew it was coming. After all, the sparkle team was created for just this purpose. As a weapons-related person, I and a lot of the personnel in the front half of the boat weren't significantly affected. For the most part, our involvement was only for drills related to the inspection. As we all hoped and expected, the boat passed and performed well. Now we will transit the remaining distance to reach port and end this patrol. One week was remaining and our primary duties are to prepare to turn the boat over to the Gold crew.

During that last week of patrol, we were busy, but not in a

challenging way. That week was difficult for different reasons. At this point on this patrol it had been more than two months without seeing the sun, experiencing fresh air or having any contact with the outside world. Plus, everyone knew that every patrol seemed to end with at least one sailor discovering his spouse or significant other was no longer to be his partner and would not be there to welcome him home. That thought weighed heavily – even to this day. Sometimes he also found his bank account was emptied.

Tasks at this point were automatic. That's important because I just couldn't concentrate – no one could. I was in a constant fog. It's almost like that initial dazed feeling when first awakening from a deep sleep. Sleep was sporadic. It is an indescribable feeling that lasts for the entire week. We all were easily distracted as the thought of seeing loved ones soon lurked just beneath the surface of our daily thoughts. I would eventually find that was true for every return from patrol, but it lifted the second I crossed the gangway to dry land. And it only happened during that last week of a patrol.

It's a sort of distraction caused by the expectation of seeing loved ones again, that I've not felt since leaving submarine duty. As the remaining days of patrol are crossed off, I made a list in my mind of what needs to happen to get Alex moved. We need an apartment, moving van, my car, furniture and supplies. Being in patrol isolation as I have been has made me question everything. I honestly don't know why, but regardless, a huge change is coming.

Remember this day – no one else will, or ought to.

Chapter 12
I Do

It's incredible to see the sun and feel its warmth. Both are free of charge and so incredibly welcoming. The sun feels like a long-overdue hug that I never want to end. It provides a magnificent comfort. I'm in a hurry to wash the submarine smell from my clothes and pack away those items only required for patrol. One obvious habit obtained at sea was to ensure everything I put away was properly stowed to prevent it falling when the sub pitched and rolled. That habit admittedly looks odd when I inadvertently carry it into the civilian world – it was just habit.

I would sometimes do a double-take at glasses sitting on the kitchen counter or find myself worried for items on shelves as if the apartment might encounter a rogue wave.

The rush I'm exhibiting is to re-introduce myself to the pace of normal life. I can't wait to eat a salad, drink a glass of real milk, and get a fast food burger. All this crowds my thinking in the first twenty-four hours of freedom. Alex and I talked last night. We are thankfully of the same mind to start a life together. To make that happen, here's the plan: move Alex.

That's it. That's the whole plan.

Okay, so it's a plan devoid of detail, but it works for us. I blame the Navy for our inability to perform long-term planning. Although had it not been for the Navy, our paths wouldn't have crossed.

Alex doesn't know of Ellen. I privately said a prayer of thanks that I returned from patrol and Alex still wants to see me. I flew to Cleveland partly for fun, but also on a mission. Seeing Alex is amazing! I stayed in her Grandma's basement again. She and I played board games on the living room floor while her grandmother sat nearby and knitted. Somehow I felt comfortable. I had no interest in anything more than that. Is it weird that normalcy seems unusual to me? We just spent the evening together playing box games. It was so relaxing – no stress anything near like what patrol entails.

Alex and I haven't had a single traditional date. Our "dates" have been the weekly telephone calls but being with her at her grandmother's house feels so good and comforting. The next night we met Alex' bother and his wife at a local bar. I had the ring I planned to give her in my left front pants pocket. I placed it there purposely so it wouldn't get caught-up in my ring of keys in my right front pocket. I'm so nervous I might lose it. I repeatedly reach down to ensure it's there.

Many months earlier I had called from among the mosquitoes to ask her grandmother about Alex's ring size. I ignorantly didn't know the technology of adjusting ring sizes is well established and sound. My defense is that the jewelry store asked me her ring size so I thought it must be important to know that. So Grandma probably knew I had the ring but didn't lead-on to Alex.

I have to give Alex this ring to alleviate this stress. I had trouble even carrying on a conversation due to the nervousness of thinking I might lose it. I didn't have much income at the time and this ring wasn't expensive as rings go, but to me it was irreplaceable. Certainly financially speaking it was.

While her brother and wife were away at the same time from their barstools in the bathroom, I took my opportunity. It was a spontaneous thought. I slipped my hand in my pocket, secured the ring, and handed it to Alex asking, "Do you want to get married?"

It was awkward, unrehearsed, surprising, grossly unusual, and somewhat unbelievable to anyone I relate the story to, but it happened. We got engaged in a dimly lit semi-crowded bar with music playing slightly too loud. She took the ring with glee, we hugged and never looked back. Just like that – we were engaged.

I feel like everyone has a beautiful or quirky story to tell of their engagement, except us. I accept responsibility for that. Our plan is a little more defined now. I have only a handful of days to implement the plan since I need to check-in weekly, in person, on base even though I'm on R&R. We decided to rent a U-Haul, pack her belongings and drive to Cincinnati, pickup my stuff and get my car, and then drive to Charleston, South Carolina to our first apartment together.

We first "shopped" in her grandmother's basement. There we were supplied with everything we needed for the kitchen, as well as bed linen and general apartment living items. Then we drove the U-Haul to her parents for more furniture and a little more kitchen-ware. Then, after an extended goodbye, we were on our way – both with nervous excitement. It was almost like that feeling when leaving on vacation. There was so much anticipation, excitement and happiness.

Looking back now, it's crazy to think how we started life together. Five hours after leaving Cleveland, we arrived in Cincinnati. There we loaded more furniture and supplies in the U-Haul, then filled the U-Haul and Dodge Colt with gas and started the long part of this adventure to Charleston.

Saying goodbye to our respective families as a couple was new. Though no one uttered any doubt about us, surely they had some. We were starting our life as a couple pretty atypically. Actually getting married still needed to be planned. It turned out to be an experience I couldn't have predicted.

The drive to Charleston from Cincinnati doesn't usually require three days, but that's how long it took us. I drove the U-Haul and Alex the Colt. The hours of driving weren't difficult. We stopped as needed and often to eat and get gas. The motels we could afford were unimpressive, to put it mildly. I told myself not to think about the possible previous users of this room and bed. Nor did I breach the topic with Alex.

The three days of travel were needed because we didn't have an apartment or place to live in Charleston. No one I know would have taken this risk. It was very risky or perhaps just plain stupid, but we just knew everything would work out. We really had very few other options. "Winging it", I guess, is our style.

I checked with a few apartment complexes before originally leaving Charleston for Cleveland. All of them recommended I call back frequently for an opening. I'm guessing they didn't think I might be calling more than once daily, but one particular complex was expecting a one-bedroom unit to be available in the days to come. Alex and I had no plan if nothing developed. Thus, we took our time driving and hoped for the best.

I called each apartment complex every time we stopped. As it turned out, we were blessed with luck. That one-bedroom unit did become available. The complex was home to low income and government subsidized tenants. That described us, low income. We made it to Charleston before noon one day to sign the rental agreement and unload the U-Haul so it could be returned. The apartment was a first-floor unit, making moving in easier, but I had honestly hoped for the second floor feeling it would be more secure from possible break-ins, but that choice was not to be.

It was just Alex and I to unload what we had much help to load. She was amazing. Over the following days and weeks, we settled in. I bought small curtain rods to jam into the windows to prevent break-ins. The first floor unit gave me great concern for safety when I was gone at sea. R&R went by quickly and then as is customary for the on-shore period, I attended classes. During one weekend, we married!

We had made arrangements to have a justice-of-the-peace perform the ceremony and it was in his home (his idea). Greg, a good friend to both of us, was the witness and I suppose also filled the roles of best man and maid of honor. Alex and I thought it humorous that part of the deal with the justice-of-the-peace is we also get one free vow renewal ceremony at a time of our choosing in the future.

On the way to getting married the three of us ate at McDonald's – our rehearsal dinner (breakfast actually). I was nervous – justifiably I thought, but Alex and Greg made fun of me. Alex wore a dress, but not a flowing white one typical of this event. I was in my dress uniform. We bought a small cake at Kroger on the way back to our apartment after the ceremony and Alex and I enjoyed it while watching football at our apartment. We are definitely unconventional.

The on-shore time is quickly disappearing. As FTBs always do, we had a week of practicing missile launch scenarios. As the time is slipping away, I'm gradually becoming more anxious. Alex hasn't experienced the separation yet but understands the requirements. We met with a bank representative for recommendations regarding what little money we had. A checking and savings account was all we needed (or had or could afford). We were poor enough to qualify for food stamps but elected not to pursue it. The representative did recommend we maintain separate accounts. He knew, as I observed, every patrol ends with at least one marriage being over and bank accounts sometimes emptied.

In fact, the Navy mandated the entire crew watch a video about the high rate of divorce of greater than 50% in the military with money problems being the primary reason, but we decided against separate accounts. Alex would manage our money. At the time, we had very little to manage. The costs associated with moving and renting the apartment drained us. Alex was basically alone in a strange city. She had a job, but little support and no family.

Patrol Three and beyond

Moving Alex and getting married was a whirlwind. We somehow figured it all out. We made some friends and I was glad that Alex would have at least a few people to turn to when I was gone. Marriage is tough under these circumstances. We often found ourselves saying to each other, "We'll figure it out". And that we did. For weeks and months that turned into years, that's how we made it. Maybe that's what makes a great marriage – teamwork.

We were so distracted with living and overcoming obstacles, marital problems never had time to develop. As always happens, it's time to take the boat over again from the gold crew. And as always, the next month of port and starboard duty will be exhausting. I now knew what to expect and how to prepare. Each day slowly increases the stress as patrol nears.

On the boat, we were quite busy preparing for the upcoming patrol when an unusual drill went down. It was a security violation drill. We've done this so many times, everyone could usually simply react, but not today. This drill was different from anything we ever attempted. We were in port but asked to pretend we were at sea pretending to be in port. That confused even the most prepared person. Why were we pretending to be in port when we were actually in port? Who thought of this? What are they attempting to figure out?

The drill didn't go well at all and we never attempted it again. It was honestly laughable. For me, going to and from sea occupied two and a half years. During my third patrol I applied to OCS (Officer Candidate School). I was encouraged to do so by some officers on board after they learned I had a bachelor's degree. One of the officers helped me with my application. I did it out of respect for those officers, but never felt any real drive in pursuing a career in the Navy as an officer or otherwise. After all, that's really what that commitment entailed.

Becoming an officer would have required additional years of Naval service to the point that staying for twenty years and retiring made sense, but neither Alex nor I wanted a military career. We looked forward to more stability, perhaps a more normal existence. I complied with applying because I felt it my responsibility to do so – I couldn't say no. If I was eventually offered OCS, Alex and I would make the decision then, but we know our initial leanings.

Patrols eventually become easier. At least I can say I guess I got to the point where the routine was comfortable and I knew what to expect. I was a MCC supervisor and Ted was my technician. He was quite knowledgeable regarding the book knowledge, but slow to react in real-life or unforeseen situations. He never really developed the ability to react quickly in stressful situations and make quick decisions. When the boat was in position and covering targets (ready to launch at any moment), we sometimes received training messages directing us to launch. The launch messages and therefore the scenarios varied and we were evaluated as to our performance. Rather than Ted jumping into action, he would just freeze for a moment, so I would direct him. This happened in fire drills, flooding drills and so on, but I grew accustomed to directing his actions so we developed a good relationship. Actually, he knew what actions were necessary. I believe he just didn't have the self-confidence to rely on his natural and instinctive reactions.

On one particular day at sea Ted and I had the watch from midnight to 6am. I started with the usual trip to navigation, next to the Conn, to understand the plan for where the boat is heading. That takes only a few minutes. From there I might swing through the missile compartment and then go to MCC to get report. There's nothing happening and nothing expected. Occasionally I would instead call the missile compartment person on watch to see if anything is happening there. Tonight I did just that. Nothing is going on there either. Thankfully that's the norm. Hopefully this'll be a boring six hours.

Verifying the missile targeting with the off going supervisor is a quick routine. So Ted and I settled into our tan faux leather captain's chairs, with dual armrests and high backs, I might add. He had some preventive maintenance routines (PMs) to complete on the computer launch system. I had some collateral duty work I could do, but I'm not feeling motivated. Six hours can become quite long if bored so I might resort to some busy work. For now though, my feet are up and I'm relaxed. These six hours proceeded as hoped, boring as hell. Ted and I talked about our families a little.

I can't indulge in that conversation for long though because it'll inevitably result in my thoughts getting lost on that topic. I'll realize how much I miss home and how out of touch I am. There's too much time left in patrol to let home occupy my brain to that extent. Ken relieved me at 6am. He is always punctual – I appreciate that. I was glad to give him a boring report and headed for my rack for what I expected to be around three hours of sleep. I'll be awakened by some drill around 0900 hours. The usual is a fire or flooding drill, but there is also the occasional security violation or other emergencies to practice as well. Usually there are two drills, each lasts perhaps a couple of hours. The drill itself, followed immediately by clean-up and re-stowing gear afterwards as needed while the drill leaders perform a debrief and analysis.

After all this, I eat lunch and relax in MCC for a while or perhaps catch a shower. I always try to time my shower when I'm confident a drill isn't about to start. Although, truthfully, a submarine shower that can't have more than fifteen seconds of running water doesn't take long even in its entirety.

Additionally, my collateral duties usually need some

attention. Among other things, I'm responsible for ensuring all documentation in MCC and the missile compartment is up to date with changes. I'm also responsible for documenting every battle stations missile, creating a timeline to the second. The WEAPS needs that for analysis as part of his report to the captain. Steve has the MCC supervisor watch now. He relieved Ken at 1200 and I'll be back at 1800 to relieve Steve. This routine is difficult because we work on an eighteen-hour day so sleep is difficult to get.

Around 1600 the boat gets a radio message directing us to 'retarget and shoot'. It is a missile launch drill from Washington D.C. We can get these any time of day. The scenarios vary widely. We were actively covering targets meaning we were ready to launch at a moment's notice if directed to do so. Soon enough battle stations missile is announced overhead. Steve is a step ahead because in MCC we monitor comms (communications) between the radio room and the OOD (officer of the deck). So we initiate the needed actions as soon as the message is received. Everyone on the boat scrambles. People leapt from their racks and slid into their poopie suits, zipping them up as they ran. Every part of the boat is affected. Everyone has a role to play.

I believe the entire boat was in place in less than fifteen minutes. In MCC is the FTBs and the WEAPS. We accomplish the necessary retargeting and verify it. We're ready to shoot quickly. All of it done quickly and then the boat moves to hide. Launching a missile is extremely noisy so we'd become a target ourselves quite quickly if we didn't take action. We never 'fire' a torpedo or missile. We 'shoot' torpedos and 'launch' missiles.

The word fire is used only in relation to a fire drill or actual *fire*. That avoids miscommunication. The rush to shoot is accomplished with a fast, yet calm demeanor. Unusual or unexpected events are handled with the same determined attitude. It is a performance of well-rehearsed actions.

However, there was this one time where a newbie (new to

the boat) jumped from his rack and rushed to MCC so quickly, he didn't feel he should take the time to don his poopie suit. He arrived, dressed in only underwear, t-shirt, and tennis shoes. The rest of us smiled, shook our heads in disbelief, and strongly suggested he return to his rack and get dressed. His intention to respond quickly was a little misguided.

A subsystem of the computer system recorded every computer action. The boat gets into position and we "launch" two missiles using a training mode. The system is returned to standby from training. All appropriate targeting is reverified. I gather all documentation to create my report for WEAPS. Everything seemingly went well from our perspective; although the captain is the judge of that. By the time it's all over, I barely have time for dinner before assuming the watch again. This has been a pretty normal day.

I only got a few hours of sleep, but it was busy so this day passed quickly. In relieving Steve I'm hoping for six boring hours again, but this time I have an unexpected battle stations to review. Ted will do the usual PMs. This is roughly how each day passes. Sometimes something breaks in MCC so our time and attention quickly switches to troubleshooting and fixing the problem while also developing a work-around in case we need to launch, but there was never a time in my memory when that wasn't accomplished expeditiously. The drills we perform vary and we get the occasional surface contact to follow, listen to and evaluate. I didn't always know the identification of the contact.

Our plan is to be unobserved by everyone, including American sea or aircraft. We stay silent and unobserved while performing what's demanded of us for a little over two months or so (it varies), then return to port. The gold crew takes over and does it all again. However, as a ballistic missile submarine, we weren't expected to perform like what's depicted in movies. Our duty was to go to sea, stay silent and invisible, and make ourselves ready to launch as directed at a moment's notice. We hid from everyone quite well.

It wasn't boring, but certainly wasn't a constant battle dealing with some enemy. There were the occasional surface contacts to deal with-we might track or follow them for a limited time before we disappeared again. No surface contact ever knew we were there. We did our job, then silently moved on. That wasn't harrowing or overly tense for me. During those instances, the protocol was to rig for ultra-quiet. That meant anyone not required must go to their bunk to prevent any noise being generated from the crew being up and about. Doors were shut, un-necessary lighting turned off and no overhead announcements allowed. If I happened to be on watch in MCC, I and my MCC tech would sit and listen. Those of us in MCC really had no significant role or action to take. Just to sit and wait it out, but I did have a certain uneasiness knowing something was out there, not knowing exactly what or who it was, but hoping they didn't detect us. I really wasn't in a position to know what kind of contact was noted. As long as our mission wasn't compromised, I didn't care much.

There was one instance of unusual need, but not involving any other ship or contact – a shipmate developed a collapsed lung. I wasn't medically trained at the time so it did not seem a medical emergency to me at the time as missile control supervisor. I had no involvement or, as I said, medical knowledge to understand the implications. The affected sailor was resting. Doc cared for the individual to the extent he could as preparations were made to have him airlifted and evacuated as soon as possible. This required careful and meticulous planning since we obviously had to surface.

Getting the sailor 'med-evaced' (medically evacuated) quickly was the priority, but it was also important not to sacrifice our position. We never would normally surface, but obviously were required to do so to make contact with the helicopter being sent. This required surfacing during daylight hours. Our normal patrol responsibilities were interrupted.

Seas were not calm, but not terrible either. Somewhat larger waves rolled through occasionally causing us to rise and fall in a sort of cork-like manner. The helicopter hovered above. It was an anxious time topside.

Overcoming the problems with predicting swells in the waves and how they'd affect the boat's movement while ensuring the sailor didn't swing into the sail of the boat while also being hoisted up to the helicopter put us all on edge, but he had to get to the helicopter. Thankfully, the efforts were successful and the helicopter safely flew away with him on board. Quickly we returned below deck and submerged.

We (the crew) didn't receive word that his medical outcome was good until we reached shore and patrol ended, but we expected him back for future patrols. That news was a big relief for us and I'm sure his family.

Alex has an impressive work ethic. She worked at a local grocery store in a position that was well below her talent level, but it was a start and we needed the extra income. Eventually she was recognized as someone management relied on for assistance. She did it all, at work and home, stepping in to perform as a supervisor at work and became the reason we saved any money at home. I was either at sea or had other responsibilities related to the Navy so I wasn't home a lot. I was at sea when I eventually received word that I wasn't accepted into OCS. The captain personally came down to MCC to inform me.

I appreciated that, and I was internally thrilled to hear that, though I stayed stoic as the captain informed me. He offered comfort saying that often the first application is turned down to see if the applicant is persistent. I, of course, would not be, but I accepted the news and let the captain provide reassurance. That news let Alex and I off the hook.

We allowed ourselves to start looking toward the future with me as a civilian, but even that is more than a couple of years away. We endured the first patrol as a couple. I finally came home and had someone there to greet me. As I and others stepped off the boat, I saw the welcome party of wives, significant others and/or girlfriends. They ran towards us en masse. I saw Alex at the rear of the group but couldn't get to her. Stepping to the side, the rush of women and some kids hurried past me allowing me to reach Alex and get the hug I had been anticipating for weeks.

Maybe being at sea is more difficult than I might admit as I feel a rush of relief and peace upon our return. We endured several patrols as a couple. None of the patrols had dramatic experiences, but certainly had some unique situations. As is probably assumed or perhaps obvious, a submarine operates best submerged. Being on the surface in calm waters is fine, but only small waves can make it bob like a cork. The unpredictable cork-like motion is annoying and nauseating to some so I much preferred to be submerged – everyone did. Besides that feeling, we all were more comfortable underwater for other reasons. That's really how the ship was designed to operate and we all felt any defensive or offensive actions were best accomplished submerged.

There was this time we were sailing out to our assigned area through — or more accurately stated, under, the outer bands of a hurricane. Its position and direction were known, but also known was that we would meet a part of it. We found ourselves altering our course, but still passed through the northern third of the storm. The plan was to pass through it and drive on. That seemed simple enough to me. The wind and waves were quite significant, to put it mildly. We were submerged, but on the back side of each wave above us, there seemed to be a sort-of vacuum that tugged at us causing us to "bob" slightly, even though we were submerged.

It was really amazing the depths to which a hurricane's effects are felt. We naturally continued to deeper depths until the storm's effects were minimized. Thank goodness we had that option. So this amazing storm of mother nature's creation was of little consequence to us. I strongly prefer my submarine experience to that on a surface ship. The only problem that would be of concern is destruction of the port where we intended to return or the effects to family and loved ones. That didn't occur this time.

It was during another patrol that we were surprised by a radio message indicating we are to be extended at sea for a surprise weapons system inspection. We had been comfortably cruising toward home. The end was near and we all had shifted our focus to family. We all had that expected head-fog that only presented itself during the last week of patrol. In only a couple days I'd be in my usual spot topside as we travelled upriver to the tender. That changed with this message. The inspection was of the weapons departments, missile and torpedo, and caused the patrol to be extended by a few days.

We surfaced to meet the small boat that would transfer the inspection team on board. I was topside to help. The usual meeting before going topside had a particular focus knowing that we were being watched. We surfaced, then came to a halt as the boat came alongside. I could see Charleston on the horizon. Home was so damn close! It was too far to say I could realistically swim there, but at this point I might give it a try had it been offered.

The inspectors came aboard, the small boat left and we all went below deck – we then submerged for the impending inspection. It lasted three days and involved battle station drills, security violation drills, watching us perform preventive maintenance routines and personal interviews. It was an exhaustive inspection. It was especially difficult since it was unannounced, unexpected, and was at the end of our patrol when our minds had already shifted to family. Redirecting our focus was very, very difficult but we performed well. When it was eventually completed, we sailed with our inspectors to port.

That patrol turned out to be my longest. If memory serves,

I think it was seventy-eight days. My last patrol included a port call. Ballistic missile submarines get very few port calls. This is the only one we got in my two and a half years on the boat. We got three days in a coastal English town. The actual time on shore for each person was two days to allow for everyone to get their chance and make memories.

We were advised against using the train system to visit other cities because of our lack of knowledge of the train system. The worry was that some might miss a connection and then miss or be late for the departure of the boat. Some shipmates ignored that advice and were late returning. That caused some other shipmates to be late getting to shore for their opportunity. The few that were late were admonished. I don't know their punishment.

We surfaced slowly, hardly making waves, and very slowly cruised across the ocean's surface looking for the English contact scheduled to meet us. The fog was extremely thick, giving us only perhaps a thousand feet or so of visibility in any direction. We were encapsulated by it. A shadow appeared in front of us in the distance and slightly to the port (left) side at about the 10 o'clock position. It approached slowly. As the shadow grew larger and darker, we finally were able to make it out. It was the English tugboat we were expecting.

We yelled greetings and it was given permission to come alongside. We proceeded into the bay of a hidden English city with tugboat assistance and guidance. As a member of the crew topside to handle lines, I stood ready to receive the monkey fist. The lines used to tie-up to subs or ships are quite large-around a few inches think, so it's impossible to throw it to another ship. Instead, one end of a small line is secured to the thick line and a "monkey fist" is created at the other end of the small line. A monkey fist looks like a very large knot, about the size of a small baseball. One would throw the monkey fist to a ship alongside so the receiving crew can eventually pull across the small and then the large line to secure the boat.

An English crew member of the tugboat was staring at us with loops of small line in his left hand and the monkey fist on his right. His stare was at personnel well to my right (his left). He raised his right arm and launched the monkey fist. He appeared to be looking in a far different direction than where the monkey fist went. He looked left and threw it right. It came to us unexpectedly like a missile or hard-hit line drive, flying with unexpected speed between mine and a shipmate's head. Thankfully, I didn't try to catch it and luckily we were paying enough attention to avoid it hitting us in the face. It had enough force to knock us flat had it hit us in the head. Its momentum carried it across the sub without touching it, splashing in the ocean on the other side. I grabbed the small line, pulling it until grasping the large line which we tied-off to the submarine. We then repeated this same process with other lines. The fog gradually lifted over a few hours to reveal the city and countryside. The slow lifting of that surface cloud, revealing amazing countryside and city, was a beautiful process.

I heard we had to anchor in the harbor rather in the harbor due to us being nuclear. I wasn't sure if that referred to us being nuclear powered or having nuclear missile capability or if that was completely false. In any case, a tugboat was used to transfer submarine personnel to and from our boat and land. Ken and I travelled together on shore for our allotted two days. He and I got a hotel room, then walked the town. The town had two obvious halves. The old city that either survived or didn't experience WWII bombing and the newer portion that was rebuilt. We stopped in various establishments to shop, eat, and drink a little.

I wish he hadn't done it, but Ken ordered a bloody Mary in one bar. The bartender gave us a bad look. We were cautioned about this. The word 'bloody' carries different connotations in England. Ken described the drink he wanted and the bartender mixed it, but we got some bad looks from patrons. We were asked and honestly answered that, yes, we were from "that..." submarine. I did not feel welcome in the establishment, but that was due to us, not them.

We hurriedly finished our drinks and moved on. Ken asked that I keep the "bloody mary" experience just between the two of us. He said he really wasn't supposed to drink alcohol as a Mormon. I didn't know that but held it secret (from even Alex) until now. I think enough years have passed.

I don't know the many miles we walked but decided to return to the hotel for a nap. Our body clocks were a mess since we operated under an eighteen-hour day at sea. I couldn't sleep. We were in England, for God's sake! And I didn't want to waste the opportunity sleeping. I got dressed and left on a journey while Ken slept. I honestly should never have done it – we were cautioned against going out alone. It was a somewhat spontaneous decision.

Without a map or knowledge of the area, I started walking toward the first English person I could find. I had met an English sailor earlier in town who was receptive to trading our submarine dolphins for souvenirs. He didn't have it on him as he was in civilian clothes, but he'd be standing guard at a nearby military base in a few hours and he welcomed me stopping by.

In uniform, I walked and continued to ask directions from everyone I saw. Since everyone gave me the same directions, I was confident I was on the right path. All the residents I encountered on this trek welcomed me to their city. Everyone was incredibly nice and receptive to me asking for help. I even had short discussions with a few of them. Eventually, I saw what appeared to be two people guarding an entrance to a possible military base. I approached slowly – they stood and both moved to greet me. I explained my reason for being there and answered several questions.

Luckily, I remembered the name of the British service member I met. As he said he would be, he was on duty, but they couldn't let me on base. I understood and turned to walk away. I knew ahead of time this could become a wasted trip, but one of those sentries stopped me and offered the same trade. I enthusiastically agreed. We each unpinned our dolphins and made the exchange. Those British dolphins became my souvenir of this port call. Ken had been awake for about half an hour by the time I returned from what was my hour and a half journey.

Next up, regret. I needed a rest from my long walk and Ken needed time to clear the sleep from his head. We decided on a small cafe for food and drink (non-alcoholic). Everywhere we went, we stood out – even though we now were sporting civilian clothes. Perhaps it was our hair, but I kind-of doubt it. As a side note though, on a submarine, there is no barber. Meaning someone volunteers to cut hair as needed. It never looked good, but with a hat, we were passable. Obviously our manner of speech confirmed us as not of this area, but even before talking, I suspect everyone judged us.

We hadn't been in the sun for a couple months, we probably looked like tourists and since the local news announced a nuclear sub was visiting, the locals gave us a suspicious eye.

It was easy to find good food everywhere we went. After the cafe, we walked to see the WWII memorial. That was moving. After Ken's error ordering the bloody Mary, we were low key, sitting in the back of cafes or out of the way in bars. Our thought was to strive to be un-noticed. I honestly don't remember, but think we were in port on a weekend. That guess is because of the crowds present in bars in the evening. That is vivid in memory. We walked one avenue and stopped in a couple bars. The first was so crowded, we just left without attempting a drink. The second was less crowded so we entered. We each got a cold beer and sat off to the side near the door we entered. As we drank, more people filled the establishment. We agreed to just the one drink and leave to get a good night's sleep in the hotel.

Between us and the door was what became an overcrowded space. Patrons no longer had any space between them – most standing shoulder to shoulder. The loud music became eclipsed by crowd noise. Ken and I started feeling nervous and thought it best to leave. Halfway to the door a fight broke-out. We weren't involved except for an errant punch hitting my face, knocking my glasses to the floor. I immediately bent over to retrieve them. I'm nearly blind without them. Thankfully they weren't damaged. Following that we pushed harder to reach the door.

Once outside, we looked at each other in relief having made it. As we quickly walked away, we observed local police going in that bar – a very quick response. They must have been near, but we hadn't noticed them. Thankfully we avoided an international incident (in our minds). The next morning, we returned to the boat, agreeing to keep our incident to ourselves. We never should have gone in I suppose.

Differences between leaving and returning

The days and weeks leading to succeeding patrols are getting more difficult psychologically. Each patrol seemingly harder. I'm becoming more cognizant of the effects patrol after patrol can have. I see people who have been going to sea repeatedly for many years. They are changing over time and I think the change is permanent. They just don't seem the same person to me. I can't put my finger on it, but I think being disconnected from humanity for so long matters, but I admit to no training or education in judging people on a psychologic basis. I never considered or was just plain ignorant of that happening.

I now am certain I'm leaving the Navy after my enlistment is over. As patrols passed, so did two and a half years. Finally, would come the time for my last patrol. After this, our plan was for me to transfer to the tender to work. It is the repair ship the submarines tie-up to when returning to port. It rarely went to sea and didn't do so for long patrols. So I'd finally have a more regular schedule and be with Alex. It's still not the normalcy that civilian jobs offer, but it's an improvement. My pay would be significantly less though.

For me, the week before leaving for patrol was a week of undocumented checklists. For her, it was one of mounting stress and increasing emotional weight. We've both been through this before. My mind was constantly shifting focus to each priority. The importance of each item on my mind's list grew each hour of each day. Playing over and over again was me talking to me: Thank God this time is the last...Is all my stuff ready to go...Will she be okay...Will she even be there when I return...If I let it, my mind could easily foresee the worst of outcomes. So much seems out of my control. For now, I am a part of something.

Once at sea, it's as though I never existed. Everything moves on without needing me. While on patrol, returning home was something I tried not to contemplate, but of course she was in my thoughts constantly. It was difficult to have these competing ideas. Keeping busy was the key. The captain did his part to keep us from boredom by scheduling many, many drills covering all sorts of scenarios. And I can admit boredom wasn't present, save for that last week used to transport us back home.

I mentioned previously that during that week, we were busy. That week was difficult for different reasons though. Knowing home was getting closer, knowing I'd see her soon, knowing I would never have to endure this again clouded my thinking. At this point it had been over two months if my memory is accurate. I've repeated that sentiment repeatedly but missing the sun and fresh air is a real thing.

Every patrol ended with at least one sailor discovering his spouse or significant other was no longer to be his partner and would not be there to welcome him home. That was quite a depressing realism. Each day at work before shipping out requires focus. I'm good at that. For both good and bad, when I focus, everything around me becomes a blur to my eyes and background noise to my ears. Small, singular objectives is how I think I'll get through the week, but home is different. I almost don't want to work on anything. It's as if I have come to believe the week can't end and I can't leave if everything isn't done. That is of course ridiculous. I'm leaving – that's the bottom line. Not because I want to, but because of choices and a commitment I made in the preceding years. It'll work out. At least I hope so.

I've always clung to hope. It has been what I've always

found to be most dependable. Each long, hard day at work leaves me exhausted. At the end of each day, I have only the energy to sit with her, watch meaningless television and turn-in early. Each day during that week on the boat of slowly returning home was identical. Each person had preventive maintenance tasks to complete. They were not many in number so we each developed our own schedule so that each day gave us something to accomplish. Interspersed among the tasks were the responsibilities each person had within their role (their collateral duties). And beyond this work was ensuring everything was in order for a smooth turnover to the next crew. Each day involved completing the needed tasks and checklists, eat at the appropriate and designated time and attempt sleep for the time left. To a person we were incredibly tired upon arriving in port, but not due to work, but rather due to lack of sleep. The time given us to sleep was adequate. It was instead the head fog that made work arborous and prevented sleep.

That indescribable fog had no redeeming value but was unfortunately undeniable and unpreventable. That morning I was to leave was routine only for the alarm clock – I was up the same time as each of the preceding days. She was too, for today she was driving me to work. She'd return, and live for an extended period unknown to her, alone. I never told her when I'd be returning or where our patrols took us.

"This is the last time," I said. I smiled to provide reassurance. As she drove away, I walked slowly, carrying what was the last of the remaining items on my checklist. Water welled in my eyes. I walked slowly and fought it so I wouldn't be seen boarding my submarine with tears. I had the same thoughts as each of my previous walks before leaving on patrol. Somehow, that walk transforms me from a living person to someone non-existent – or certainly to no more than a memory. I felt it was as if I no longer existed. Everything moved on. Nothing and no one needed or depended on me. It's not as if I died though. At least that would have meant I existed.

This walk somehow erased everything, or at least that was the assumption in my mind.

In some way, perhaps, that meant I had nothing to lose. If I didn't return, nothing would be different. Thankfully, she was quite capable. I've always said, that's what I love most about her – she is independently dependent. She would be fine without me, but still wants me around. Just as crossing the gangway voided my life, standing topside when returning gave me life. It is not unreasonable to call it a re-birth of sorts. I was feeling renewed while squinting as sunshine warmed me. My lungs filled with an unusual feeling of freshness. People boating, surprised at our unexpected appearance, stood, smiled and waved. It is an unintended pun, but I felt buoyant.

I was for the first time in many weeks, relaxed as we navigated up the river. Our destination was the pier or tender where family had been gathering for hours. Wives with kids balanced the anxiety of finally seeing their returning love while controlling the children who had no real understanding why they were being asked to practice control for a seemingly endless amount of time. Of course there was always a few spouses without children who welcomed their returning partner with nothing more than a long coat and nothing underneath. And then at least one buoyant returnee was met with a gut-punch. His significant other and perhaps his bank account was gone.

I had volunteered for the Navy and submarine duty. I made a commitment. Commitment in work and love was a reason for living. Had I not that, certainly running away was understandable and plausible. The gangway is now in sight. It would transfer me to that void in my headspace and the submarine in reality's moment. It made me unnecessary to everyone I'd ever met. It made me non-existent. If I were to return, it wouldn't be due to the passage of time. For in this void, the passage of that wasn't measured, felt or tended to. I would return only after achieving the goal or meeting the standard.

Meeting some goal would haunt me the rest of my life after the Navy as well. I don't know how she felt in these times. I'm not sure knowing helps anything, but she, like me, had no comfortable way to say goodbye. That hug upon return is indelible in my memory. It made me whole again. I do love a good hug. To this day it gives me incredible energy somehow.

Returning from patrol, I feel I'm a stranger – not just to the sun and fresh air, but to all of human-kind. The blinding sun was amazing, not an annoyance. Birds and bugs were beautiful. At home nothing slid around in cabinets or off of bookshelves; although it took time after each patrol for me to relearn that. It turns out, apartments don't pitch and roll. I unfortunately stood out due to an obvious lack of sun.

My bed at home wasn't cramped and hallways were wide enough for two people to easily pass! I'm finally home at last. I missed her so very, very much, but I wasn't one to verbalize emotions and since that experience with Ellen, try hard not to show emotion. Looking back now, I have no idea how we did it — but it was certainly a 'we' thing — but we are good at change, adjustment and modifications or at accomplishing anything put before us. We are really, really good together. I don't doubt us.

Many years and challenges laid before us at that time. We had no idea what to expect. We didn't care or think about it. That was a great plan, as it turned out. I know some military units have special insignia they wear on the uniform sleeves that represents them. I and others who did my job on the submarine had one too, but it wasn't worn on our uniform, but instead displayed as the first page of our military record. It was simply a unilateral triangle with the following words on the sides: Integrity – Honesty – Trustworthy.

Now I see why God created the waters,
for a man who wants to be free.

Chapter 13

Making Repairs

A new job and family member…

The tender is much larger than a submarine, obviously. I'd estimate it's approximately the size of a small cruise ship. Perhaps one that can accommodate three hundred or so passengers. Those of us working on the tender and also submarine qualified are a minority of the crew – not a small minority though. The ship has two cranes on top, but otherwise quite non-descript. The aft or back portion has an open area that could accommodate a helicopter landing if necessary.

I passed two guards at the entrance to the long, wide cement pier and continued walking to the end where the tender was docked. Its width is enough for two cars to easily pass though that type of traffic would be a rare occurrence; however, trucks of various sizes do frequent the pier. The foot traffic on the pier is constant, but not busy. The busy times are in the morning with sailors arriving, around the lunch hour and at the end of the workday. Accompanying the pedestrian traffic are forklifts and small people carriers similar to golf carts. I've done this walk plenty of times before, but always continued through and across the ship to a submarine. That's not the case today.

On the tender's long gangway was a line of sailors, enlisted and officers, waiting to be allowed access. When it was my turn, I saluted and requested permission to board from the sentry, showed my identification, and notified him of why I was there. He had me step aside while someone retrieved a member of the quality assurance department to provide escort. My role is to eventually be the primary or lead quality assurance inspector for the FTBs.

The FTBs on the tender provided technical and maintenance assistance to submarines and helped during missile transfers. My work schedule developed into basically a 9a-5p, Monday-Friday, routine. However, I was required to spend one weekend per month on the ship. The biggest downside to working on a surface ship was losing a significant amount of pay associated with submarine duty – at that time it meant losing sub pay and hazardous duty pay.

Alex and I tightened our belts, especially since our first child was due. I started in January of 1988. Our first-born arrived early in the month. My six-year tour of duty in the Navy will end in January of 1990 – two years to go until I had the option to return to the civilian world.

It was during the overnight hours when Alex' water broke. Her first response, "I don't want to go to the hospital."

My first response, "We need to get to the hospital!"

It was interesting that we had diametrically opposed initial thoughts. We both knew her water breaking meant the baby had decided it was time to become a more visible member of our family. I grabbed the two bags we had ready, one for each of us. Arriving to the hospital and checking into a room, she and I turned off the lights to nap. It was much earlier than our normal wake-up time and we were tired.

A nurse came several minutes later, turned on the lights exclaiming, "Get up, we're having a baby!"

I was in a recliner, next to Alex, she in the bed. Alex's packed bag was on the floor next to the recliner and my bag was outside of hers. My bag was simply packed. I brought only a large package of Twix bars. Alex packed sensibly for her hospital stay.

The ensuing labor lasted fourteen hours with Alex, through most of it, in a large amount of pain. She received an epidural late in the process so it had little effect in addressing her pain. I suppose it helped to some degree, but I'm confident the pain she endured was far greater than expected and much, much more than I could have tolerated.

To her immense credit she was never disagreeable. At least not so beyond the occasional short bursts of irritation. To this day I still hear how I wouldn't let her hold my hand during those obvious episodes of great pain when she was pushing. My only defense is that our Lamaze instructor told us not to allow it as she would probably crush my hand during contractions. I was following the recommendation given me.

I was apparently of little help and maybe even in the way as the nurse suggested I take a break in the cafeteria while we were in the middle of the fourteen-hour marathon. I did as I was told.

Towards the end of the fourteen-hour labor Alex was working really, really hard. I knew that, but to anyone who didn't know her, it was hard to tell. The nurse said, "Okay, go ahead and push."

Alex took a deep breath and pushed, but her face didn't show it. It was with a very calm look that she gave great effort – no grimacing. The lack of expression caused the nurse and I to glance at each other and then back to Alex. The nurse asked me, "Is she pushing?"

"I don't know," I responded.

Then she asked Alex, "Are you pushing?"

"YES!" said Alex, very emphatically. At that moment I knew I was in trouble for not knowing. When it all ended with our baby's birth, I was permitted to cut the cord. Our son and Alex were perfect and their ensuing hospital stay uneventful. Of note, I changed his first diaper (he peed on me).

Luckily, I had enough Twix bars for the entire fourteen-hour labor. I was proud that I estimated correctly.

Upon discharge and arriving home, we started a new chapter. It was me as completely ignorant in caring for a baby and Alex as the subject matter expert. Starting a new job and becoming a father simultaneously carried with it an increased level of stress, but it was less than that of a submarine patrol.

The new job started with many hours of training. I eventually became a qualified inspector in my field as an FTB, but also as an inspector in the torpedo department. It wasn't new, but somehow a different and welcoming experience to work in an environment that includes women. That made the job feel more normal, like a civilian experience. It also was a new experience performing inspections in submarine superstructures where alligators sometimes take solace. With certification as a quality assurance inspector also came work as an inspector ensuring safe transfer of items using the onboard cranes as well as working closely with the Marine's involving security.

I suppose it could be said I had my hands in a lot of pots, but it was all mostly satisfying and helpful in preventing monotony in the job. Although, time gave me perspective and an understanding that I definitely would seek a change when my two-year assignment on the tender was up. I didn't hate the work, perhaps it's best said that I found it tolerable. My normal routine was in observation of work by the FTBs on the tender and occasionally on subs, but I gained the most pride in the one instance of being assigned to observe and inspect the missile technicians and their department on the tender.

We all had an upcoming inspection and I was assigned as their quality inspector to prepare them for that inspection. The idea was that I could bring a new or fresh perspective. Others were assigned to do the same with the FTBs. I was tasked with inspecting their department to ready them. I approached my responsibility with the perspective of I and them working as a team to identify/correct deficiencies rather than me pointing out their faults. They were quite receptive to that and as a result, much more open. We found a lot to correct, but they did it and as a result I could report to my supervisors a well-organized department of missile technicians with very few deficiencies.

That would unfortunately be the only instance of involvement of me with them, but it was certainly rewarding. Beyond the routine tasks was monitoring torpedo-associated tasks and other related inspections.

Alex was a great mother and I a fast-learning student of fatherhood (strictly my opinion). Thankfully our son was thriving in spite of my apprenticeship. We called him Bob throughout the pregnancy and at the hospital until birth. Neither of us knew why, but that name appealed to us during that period. However, we didn't know we were having a boy until he made his appearance. Following his birth, we selected a name very different from that. Most people looked at us confused, thinking 'Bob' was our preference all along, but it never was to us. 'Bob' was the pregnancy name. We liked it, but never intended it to be his lifelong name. I suppose that was similar to everything else we did in that we're just a little different. Hopefully not in a bad way.

Alex gave me tips regarding diaper-changing so I gradually improved my technique, eliminating the problem of getting peed on. She also taught me how to feed our baby. I really enjoyed feeding and bonding with him. Life was new, in fact it was actually very different and better. Our next-door neighbors were an older couple. He was a Navy 'lifer' meaning he would put in the necessary time to retire from the Navy. He also worked on the tender, although our paths never crossed on the ship. She came over occasionally to play with our son and fill the grandmother type of role a little. Occasionally she brought food.

Neither Alex nor I were particularly fond of the taste of her efforts, but the gesture was much appreciated. All of her visits were with a grand smile and hugs. Being at home without the stress of looking forward to long patrols at sea was fantastic but going to sea wasn't eliminated. It was just for shorter periods and less often. I don't think we were ever gone for more than a week or so. I believe our trip to the Bahamas was the longest time at sea for me. That trip and my visit to a casino slot machine there yielded me $40 in winnings. I'm not inclined to gamble. I honestly don't enjoy it, but a slot machine played well on that day.

Beyond that memory, the most memorable of remembrances were the night skies at sea. They were the most beautiful I've ever seen. They were mesmerizing. The stars and moon were incredibly picturesque and their reflection on the ocean's surface was equally impressive. Qualification in various jobs and roles on the tender wasn't difficult – just time consuming. Standing guard at the end of the pier and on the tender allowing people to board were the only two occasions where I carried a gun. That is, except during a security violation drill. None were difficult jobs. The most challenging of those experiences occurred on one early morning on the pier.

I and another person, armed with shotguns, stood in the dark at the pier's entrance ready to respond to any threat. The usual pace played out. A bundle of newspapers and soon after that, several boxes of donuts were delivered before sunrise. We placed the deliveries in the guard shack as usual. We notified the tender security of those deliveries and continued our watch. A twelve-gauge shotgun was suitable to counter most threats until reinforcements arrived (meaning the Marines). That was true most days, but not today. What happened on this date is seared into memory. Maybe that explains my preoccupation with donuts to this day.

On this date, the violators were a family of raccoons on the scent of the delivered donuts. We first noticed their eyes peering from the darkness from under the pier but couldn't make out much more. Slowly they moved towards us. Their bodies and their numbers soon became evident. There were more of them than us. We both knew that firing our weapons on this threat would cause trouble (for us, as well as the raccoons). I alerted the tender to "... get here quickly. The raccoons want the donuts." All I heard back was laughing.

That left me with a lonely feeling that probably only those in a foxhole during battle could understand (hopefully my satire is evident). Reinforcements weren't coming. The threat was ours to face alone. I remained at my post, alone in the dark, outside, on the pier, still unaware of the significant threat facing my partner. At one point I turned to locate my fellow sentry. She was standing on the chair in the guard shack holding her shotgun and two raccoons up on their hind legs reaching up to her from the base of the stool. To her credit, she didn't scream.

The donuts were still safe on a shelf – too high for the raccoons to reach. I ran to the guard shack yelling and waving my gun. My ruckus apparently scared the varmints enough that they quickly retreated under the pier and into the darkness again. She bravely came down from the chair and back onto the pier, closing the door to the guard shack behind her. Only a few more minutes passed until the raccoon eyes reappeared, but also coming toward us from the tender was someone retrieving all the deliveries.

Thankfully, the allure of the donuts would soon be gone –but not their scent. We took solace that the sun would soon be rising. The raccoons never visited during daylight. Their threat would not be a danger for another twenty-four hours. We also found comfort in knowing the person from the tender retrieving the donuts didn't witness our handling of our raccoon scare.

Family very much wants us to come home so they can meet our new son. We also would like to see family again. It has been quite a while so Alex and I decided to take our first family vacation driving home. It wouldn't be a vacation in the truest sense of the word, but all our vacations (for a few years) involve returning to visit family.

I was approved for one week of TAP (time away pay). We had recently purchased a second car and decided to drive it. It was a cheap used car with vinyl seating throughout and an ant infestation in the trunk. Ridding us of the ants is a story all its own requiring much effort, but suffice to say, I did eventually eliminate them with multiple eradication efforts. In my regular cleaning of the vehicle I applied Armor-All to the seats. The car and its seats, in particular, heated quickly in the South Carolina sun. The trip home was miserably hot.

Our baby cried seemingly constantly requiring many stops for all of us to cool off. The drive seemed endless. Seeing family was nice, but overall, that vacation wasn't fun due to the drive there and back again. We never made a trip like that again, but a few family members did fly in to see us once. Picking them up at the airport in that car was very entertaining for me.

The Armor-All made the vinyl seats in the rear incredibly slippery. Our family visitors didn't use seat belts while in the backseat. That resulted in them sliding side to side when I cornered. I watched as they slid in and out of view of my rearview mirror and heard their fingernails scraping in futility against the plastic door coverings desperately searching for something to hold on to. I cried, laughing so hard. Luckily they also saw the humor in it. I never again applied anything to those seats, but honestly do not regret doing so previously.

Going to sea on the tender wasn't eventful, except for that one security violation drill I still find entertaining to recount. That event and perhaps also that time during the hurricane.

Other than proving we wouldn't sink, I never knew the overall goal of each cruise. They lasted roughly a week, but the night sky in complete darkness at sea was incredibly amazing. I'd like to see that again someday. Being away from land and the city lights allowed for viewing some beautiful skies, constellations and stars. The darkness revealed far more stars than I ever previously witnessed. This is certainly one benefit of surface ships over submarines. That and experiencing the waves at sea was fun.

The hurricane wasn't enjoyable, but the occasional light storm broke the monotony as we rode the waves. I was luckily never seasick.

To keep busy, I started the process of getting my surface ship qualification. It was a requirement without a specified timeframe for completion. Since that accomplishment held no interest for me, my plan was to stretch it out for two years. In other words, I dogged it until I got out of the Navy, but it was interesting learning the various aspects of the ship, like the engine room. I sometimes stood watch in the depths of the ship near the shaft that turned to propel us across/through the water. It was boring sitting with a fellow crew member for six hours, but also an experience civilian life doesn't normally offer. I gained an appreciation and knowledge of pump and valve operation that I would later apply in my civilian work.

One evening at sea I was relaxing with others in the quality assurance department watching some movie when a security violation drill was initiated. I ran to the small arms locker to receive a shotgun with a pistol grip (like those I used on my submarine) and was then paired with two other crew members, all of which I knew. I took this kind of stuff seriously. I figured that one of these days it might be real. We were tasked with clearing some compartments on a level just below sea level. It was one of my teammates leading the way, I followed him and then the third person was behind me. We held our shotguns at the ready to lock and load at a moment's notice. One of the drill observers accompanied us. We encountered no problems yet.

Each compartment thankfully had been evacuated, as they should be. Two compartments were left to examine. The man ahead of me cracked the door to the second from last compartment and become suddenly alarmed. He readied his shotgun (though not loaded since this is training) and yelled at the person he spied to halt. I and the other person in our group also readied our firearm instinctively. We swung the door wide open and saw what our team leader was so excited about. Some idiot was sitting on a stool in the middle of the compartment because he or shipmates had painted him into that spot. The paint on the floor wasn't dry yet so he refused to budge so as not to walk across his fresh-painted floor.

The three of us discussed options and decided it best to convince our drill observer to ignore him and move on. Us and the drill leader just shook our heads. Thank goodness this is just a drill. I learned later that he was forced to walk across the wet paint to exit, painting over his footprints as he went. His shoes were thrown away.

One of our excursions was a short trip north, up the coast to New York. It was Fleet Week–a celebration I had never heard of previously. I supposed it to be a celebration of all things military. A repair ship wasn't probably of great interest to most in the public, but New York was very much of interest to the crew. Everyone was excited to experience it. We were instructed to travel in groups and required to wear a uniform at all times. New York gave us front-of-the-line privileges at all the typical tourist attractions. I took advantage of it at the stock exchange.

Police were incredibly helpful and all very considerate and helpful. I exchanged my dolphins for a police insignia as my souvenir of the trip. The USO provided tickets to an Off-Broadway play. The visit allowed me to reconnect with a college roommate and fraternity brother. I rode the subway free of charge and spent a night at his home in Connecticut.

Everything about my New York experience was positive.

This type of cruise would never have been considered on a ballistic missile submarine. Fire drills, as with everything else, were very different on a surface ship. On a submarine, when a fire drill (or other emergency) was announced, I ran towards it. Being submarine qualified meant I knew the location of every fire extinguisher and the type so I could grab the one most appropriate for the fire I thought I was heading towards. I also knew the location of every fire hose and its length so I could select the one I knew would reach the affected area. I knew the location of EABs. I knew the ventilation system so could isolate the compartment, preventing smoke from contaminating other spaces. In short, I ran to the fire or emergency. My first fire drill on a surface ship was enlightening and educational. The overhead announcement of the fire came and I leapt into action. I didn't yet know my way to every space but jumped up from my desk and grabbed the nearest fire extinguisher to do whatever I could when I got to the fire. I immediately noticed no one around me moved. Everyone continued to sit and wait.

"It's okay, all the new submarine people do that," said one of my co-workers.

I responded, "What do you mean?"

"Sit down and wait," he continued. "The fire team will respond first and if more help is needed, then we'll go".

I did as he suggested. I returned the fire extinguisher to its location, sat, and waited. No further help was apparently necessary. I felt inadequate and uneasy. I felt like I should be doing something, anything, but a surface ship is definitely not a submarine. As I thought about it, fire and flooding on a surface ship do have different consequences I suppose. I realized I have a lot to learn.

One qualification was as Junior Officer of the Deck (JOOD). I didn't stand watch as JOOD often, thankfully. It entailed carrying a handgun as I allowed personnel to enter and exit the ship. Everyone, save commanding officers and above, were required to salute, request permission to board or exit and present their identification (ID). Officers and enlisted held those requirements. I'm not sure every JOOD was as strict, but I made no exceptions – even during morning hours where it was the cause for long lines to get on board. I was proud to have kept personnel from inadvertently leaving the ship with apparent confidential information and also proud to have caught someone testing me with trying to obtain entry without a proper ID.

It was quite stressful for me personally. I always felt to be under a watchful eye. Upon reaching the twenty-month milestone aboard the tender, I began preparing to exit the Navy, turning down the significant resigning bonus. Had I stayed in, I'd have to commit to four more years, resulting in a total of ten years enlisted. At that point, I thought I might as well stick it out until I qualified for retirement at twenty years. I'd also have to go back to submarine duty after the two years on the tender. In those days the Navy wanted people at sea. That was something I just couldn't do, psychologically. Although I was offered a signing bonus totaling in significant thousands of dollars, Alex and I decided to test civilian waters.

I mailed my resume to potential employers and interviewed via telephone. I received several job offers. This experience was far more pleasurable than that immediately following college. Alex and I were ultimately focused on two potential offers, one in Washington D.C. and the other in Columbus, Ohio, but hurricane Hugo interrupted my job search.

The effects of the hurricane affected us during the last three months of my enlistment. After the hurricane I used a pay phone on base to do interviews since our phone service was still out. The experience of that with the incessant mosquitoes brought back memories from those days calling Alex. I ultimately declined the Washington job. The pay was excellent, but the cost of living made disposable income too low in our opinions. Alex, our son, and I packed all our belongings and drove home to live with my parents in Cincinnati, Ohio while I awaited an offer regarding the Columbus job or perhaps continue my job search.

As is seemingly always our plan, we just figured it would all work out. We saved enough money to get by for a few months. It was the first week of January 1990 when I officially became a civilian again and finally got the offer that I accepted. The pay was triple what I was earning in the Navy.

We honestly felt kind-of rich. In reality, we just wouldn't be low income anymore. We had barely settled into my parent's place when we found ourselves already planning another move – this time to Columbus, Ohio. Soon our travels would stop there to raise a family. I would be two hours north of my hometown and Alex would be two hours south of hers. We'd be in the middle, in so many ways. So two weeks quickly passed, then we moved again and our new life would begin.

Everything following the Navy is a lot better than it was following college, just as I hoped it would be when joining the Navy. Sometimes a plan works out. I got the training and skills in the Navy that now, coupled with my college degree, helped me land a good job with a future, but it also left me with thoughts and memories that I can't erase or ignore.

As thoughts venture deep into the past, I am mostly thankful for a love that will last.

Chapter 14

Navy is Finally Over!

Well, I guess Dan did it. It seems the plan he developed six years ago worked.

He and I reconnected via a telephone call. We haven't seen each other in years but managed a few rare moments over short phone calls to catch up. These last six years had been a mystery to me as far as his efforts. He had disappeared. We had a few mutual friends, but no one seemed to know anything regarding Dan. He completely disconnected from one life and created another. In some ways, it's impressive to me.

I feel like I've done well but have lived just the one life. Dan rebuilt a second one. Or should I say he started over to create a second one? We talked for about an hour on the phone. That was the longest conversation we've had in quite a long time. I think I just met a new friend. He's not the person I knew previously. We grew up together experiencing many highs and lows. We related to each other then, but not anymore. Our current relationship isn't better or worse – I'd just call it new.

After serving on a submarine, he did time on a repair ship – a surface ship. I guess it services the type of submarine he was on. He didn't go to sea much on the repair ship. He describes the frequency that this surface ship goes to sea as just often enough to prove it can. That is apparently three to four times a year. The ship is nicknamed the Love Boat. Apparently the co-ed nature of its crew results in many rumored and secretive short-term relationships.

I'm sure there's more drama and stories than Dan is letting on, but his job was mostly a 9-5, Monday through Friday position, with a weekend each month requiring he spend it on the ship. He said he didn't mind going on subs to supervise repairs since he wouldn't be going to sea on it anymore. The repair ship job allowed him more time with his wife, who I haven't met, and his first child. He has a very young son now and loves being home with both of them.

I agreed with his sentiment, but really can't relate. He and I have grown distant with the Navy experience adding to the differences our lives exhibit. He states his plan is get out of the Navy when the time comes. Dan takes great pride in being part of the submarine community. Serving on a surface ship just filled time. He doesn't expand much on experiences involving either to me and I'm not comfortable pressing him on either, but those experiences and the training from the Navy have served him well.

The goal was to develop skills transferrable to civilian life. He has accomplished that. He achieved his goals and more – he has a family. His knowledge of electronics/electrical circuits and ability to read/perform computer programming coupled with a paper technology degree has made him very attractive to employers. His job seeking experience now is opposite to that which he experienced after college. Back then, he had nothing but bad luck that led, in his words, to depression or at least a feeling of being directionless and worthless. Now he has multiple job offers for him and his wife to consider.

Who would have thought that a degree in papermaking could be partnered with Navy training? His six-year investment in himself wound up being a great decision. Everything he intended, happened.

Dan, his wife and son made it through tough times. Being

in the Navy and quite poor presented challenges, but they got through it and have high hopes for a brighter future. The civilian career they eventually decided he should pursue was installing computer systems in paper mills to automate the papermaking process. It required a lot of travel around the country, but they feel they are prepared for that and well-suited to endure it. After all, if they could endure the separation required by submarine service, surely a two week business trip wouldn't be problematic.

Dan, Alex, and their son had only months left in the service. He said the Navy permitted him to use remaining vacation time to get out early. They made all the necessary arrangements to take advantage of that offer. They decided to rent another U-Haul for the return trip. Him driving it and Alex driving the small family sedan brought back memories of their trip some years previous, but this time the car would be packed full and also contain a child.

There was only one thing to contend with before leaving–a hurricane. It was just another challenge for a young marriage. The Naval six-year commitment was ending in January of 1990. In September of 1989, shortly before leaving the Navy and Charleston, South Carolina, they met Hugo. That's hurricane Hugo to the rest of the country.

It made landfall in Charleston as a direct hit in September of 1989, I believe. I referred to it previously. Of course, everyone knew it was coming well beforehand from the news reports but weren't sure where it would make landfall. Their son was a few months' shy of turning two years old and they had no family within a reasonable distance. The whole country watched the storm approach for days. It was predicted to be a devastating storm (as all hurricanes are I suppose) and one that would show itself to Charleston in incredible fashion.

Finally came the point where the Navy decided to move all ships out of the harbor and away from the hurricane's path. Danny had to board the ship, having said goodbye to his wife and son. He was on board only hours when he got word that Charleston would be a direct hit as landfall for Hugo.

He called his wife. The call was limited to literally only seconds since the telephone lines were being disconnected at any time. He told her Charleston will be a direct hit – go west. Luckily, they had a very good friend in a shipmate and his family. Danny's wife and son would evacuate with his shipmate's wife and their daughter. Danny had no idea where she would go or how they'd fare. He said he privately said a short prayer.

I'm sure everyone on board was saying goodbye to someone. It wasn't a topic needing discussion. Cell phones weren't yet available so he'd be cutoff and at sea, again. Danny's firsthand account of the hurricane experience on a surface ship left me unable to ask questions. It obviously left marks in his soul.

The ship eventually was able to sail upriver to the Navy base and dock. It was late afternoon when Dan departed the ship. A fog settled in at what was previously tree-top level. It was like the Navy base had a ceiling. His car in the large open parking lot somehow escaped damage. He left and drove down the road and through what was the forested area that surrounded the Navy base. Sunlight was limited due to that thick fog. Headlights made the scene feel ominous and spooky. The fog settled in a way that allowed for clear vision on the road, but created a ceiling of around twenty feet, making him drive under that thick fog blanket.

Every tree was snapped in half, in the same direction-both impressive and somewhat creepy. Main roads had been cleared by piling and pushing debris along the roadside, so he could make it to the duplex he and Alex rented and called home. Toppled trees were all around, but the driveway and their car parked on it was clear and somehow without damage. He was left to wonder where his family could be and how they got there since their car was at home.

Those questions unanswered were better than wondering if they lived through it. Although, them surviving the experience was only his hope and assumption since telephone service wasn't available. He assumed they used the friend's car, but he still didn't know how they fared. Anything could have happened. The storm destroyed everything in the area. Too many people had died.

Entering the duplex, he noticed everything was moved from closets and stacked on the couch and chair in the family room. Clothes, still on hangers, were piled high in no particular order. At the end of the hallway was the master bedroom. A tree branch apparently pierced the bedroom roof allowing rain to flood the bedroom. He splashed as he walked across their carpeting. The high humidity inside the duplex caused him to sweat without relief. Thankfully, someone had already patched the roof with a tarp. mosquitoes were terrible, even indoors, making comfortable sleep impossible.

His thoughts suddenly transitioned to memories of calling her at a payphone, inundated with mosquitoes.

The air conditioner was inoperable since there was no electricity. The heat and humidity was unbearable! In the kitchen, all canned goods, liquor, and anything edible was gone. It turned out, Alex didn't take them – all of it was stolen. They'll never learn who did it. The only thing left was a loaf of bread and an enormous jar of peanut butter he learned later his parents brought. There was nowhere to buy food. Grocery stores and restaurants were damaged and closed. The only restaurant open was a Waffle House, but they dramatically raised prices due to their advantage of having a monopoly. Dan and Alex would later make a pact to never eat at one again. At this point, thirty-two years later, that pact hasn't been broken.

He still had no idea what happened to his family or how to contact them. Just finding an operational telephone was difficult. Dan eventually called Alex's parents using a telephone on base and learned his wife and son rode out the storm in

Charlotte, North Carolina, sitting in a motel bathtub scared stiff with the wife and daughter of that close friend and shipmate. The hurricane followed them there. Most important to him, they were alive and well, somewhere. Somehow a fraternity bother's extended family from Charlotte, North Carolina heroically came to help the women and children. It's amazing how life happens.

Dan's parents came to their aid after that. They drove many hours (with that large container of peanut butter) to help in any way they could. Then some days later the women and kids drove to Alex's grandmother's place in Florida. All this happened with no help from Dan as he was helplessly floating in the ocean staring at the devastated beach.

Dan drove to Florida to finally meet his family. They exchanged stories of their experiences. Impressive was the unselfish assistance from so many – including many strangers. Their empathy, consideration and initiative was incredible. They are owed so much. They all inconvenienced themselves mightily to provide comfort and security to the ladies and children.

Situations like this played out all over the state. Their duplex and the entirety of Charleston was still a mess when they left. Even the basic cleanup of leveled trees will require much more time and effort. There was significant progress made, but so much more was left to do. The roof of their duplex was still sporting the tarp and mounds of broken trees and their branches still surrounded them. With the vehicles loaded, they started the trip home to Cincinnati.

As it turns out, they wouldn't return to Charleston until their kids, two sons, were older teenagers – from this point, many years in the future. Their current plan was to stay with Dan's parents until starting his new job. It would turn out to be only a couple weeks.

In the rear-view mirror was unspeakable devastation. In front of him was his family's future and a year of events none expected. He and Alex were excited for their family's future, and for good reason. The pay and benefits being offered would provide security they were not accustomed to. Dan's souvenirs of Navy life are the memories that continue to interrupt sleep though he would explain still of minor importance.

His thoughts of being willing to launch nuclear weapons have left him with a sense of guilt and that episode with his shotgun pointed at that teenage boater bother him, but he acknowledges that those remembrances are infrequent and really not of great significance in the grand scope of things. Many veterans have far worse memories so he realizes it could be much worse. His focus has always been to look forward – even if that is without a defined plan. So on he goes.

Don't look for explanation or reasons for the past.
Understand a lack of control-it can't be helped, what just didn't last.

Epilogue

Learning of Ellen

After sub school I had tried to re-connect with Ellen. I wasn't successful, but still carried her memory. My last memory of her was our stroll in the outdoor mall. I would eventually gain new insight, made up somewhat of my choosing or perhaps assumptions, in a letter from one of her family members. That letter was among a stack of junk mail my Mom gave me during my visit home following sub school. I had taken that stack of junk mail to throw it away, but in a quick glance noticed one envelope was different from the rest. It was from the same person who wrote me previously regarding Ellen.

Apparently, the Navy forwarded the letter home from boot camp after I left. I took it, left the remaining pile, and rushed to my room. A short letter was inside. She, the author of this letter, explained Ellen was eventually given my letter from boot camp and expressed happiness to hear from me and intended to eventually respond. She was doing well, they thought, but she regrets not yet having an opportunity to respond to my letters.

The letter's author wished me well and hopes Ellen will soon send her own letter. Unexpectedly, at that moment, I had a sudden, incredible pain in my heart. In my gut I felt there was something very, very wrong with Ellen. Emotion swept over me. Then it was like a sudden fog overcame me. I felt disconnected and weakened physically and lost mentally and emotionally.

After a few minutes, I sat, realizing I had lost awareness of my present condition. I guess I just lost it for a few minutes but regained my composure. I'm glad I was alone because I was embarrassed by my reaction. How could I have not known something was wrong with her? I don't like that my emotions felt out of control – even if only briefly.

I promised myself that won't ever happen again. I won't allow emotion an outlet. At some point over night I gathered myself enough to pray for her and her family but developing in the back of my mind was a concern that I jumped to an unrealistic conclusion. I really have no idea what to pray or wish for. Why was that person responding for her? My entire reaction was based on a 'gut feeling' and admittedly some confusion.

The next day I took time to call the manager of the grocery store where I worked. I felt I needed more information to hopefully gain a more definite answer. He and I had developed somewhat of a friendship. He was always appreciative of my efforts at work and I thankful for his recognition of that. We reminisced a little together and then I asked about Ellen. To preserve Ellen's privacy, he wouldn't tell me what he knew or if he knew anything at all. I asked if something serious had happened to her. He didn't deny it but didn't want to comment. So I suppose I was still without facts. However, it's really none of my business. I've said that over and over again. I keep repeating that sentiment using my inner voice. To be honest, I didn't know if knowing of any possible outcome was better than having no idea. I think that someone deciding to write me was a difficult decision. It had to be a very tough letter to write – especially not knowing me.

I never again received correspondence from Ellen or that person, but I also didn't write again, especially since Alex came into my life. I didn't tell anyone, save one person, but in trying to think positively, Ellen might be out there somewhere. Maybe she's even wondered about me a little. At that time, I didn't think I'd ever be ready for any more relationships, but of course, that was before Alex.

Keep memories of the past.
Be respectful and they'll last.
Keep friends in the shadow they cast.

A glimpse to the future...

Years passed, dotted by milestones representing by an accumulation of jobs and responsibilities. Challenges both personal and professional have passed with the time, but eventually came an unusual opportunity – a chance to begin again. Perhaps better said as the creation a new beginning, reminiscent of the decision to join the military.

This time though, I had a partner helping me. We've made a life of meeting each challenge. We went headlong into this one, just as we did in previous years with other decisions. This was a decision made quickly – in only a few months. Can that really be described as quickly? I'm honestly unsure, but the context is that of a thought process involving many years of casual consideration. It was finally the time.

Kids were old enough to allow for guilt-free time away. My career no longer involved travel and the economy was decidedly worsening towards recession without a preferred resolution on the horizon. A lack of bold action on my part would be regrettable. My first degree from twenty years previous provided academic credit that was transferable. Everything just fell into place. The barriers to obtaining my next degree were significant:

- All classes were at night after long days at work sweating in the paper mill and on-call seven days a week, 365 days a year
- The campus was a forty-five-minute drive from home
- The cost was entirely mine to bear
- I felt it necessary to conceal my efforts from my current employer.

But I was going to do it, finally. Alex and I agreed it was time. In all honesty, I felt it was now or never. I felt it to be somewhat similar to joining the Navy all those years ago. Actually, this decision is something I probably should have made twenty-five years previous to this day, but never mind that, it's time for a major life change, for the second or third time I suppose but that's a story for another time.

I'm not asking for agreement in anything I do.
Just your love so I can do what I feel I must do.

Teenager writing assignment result

Give an answer to how, know something of why,
Be careful with who, but comment only for illusion or a wish to comply.

Be a true friend, always lend an ear.
Don't talk in opinions, listen and consider-then alone, you'll know fear.

I'm not asking for agreement in anything I do.
Just your love so I can do what I feel I must do.
Lonely gazes and secrets with time become free
Now if only responsibility I could concede

Don't look for explanation or reasons for the past.
Understand a lack of control-it can't be helped, what just didn't last.

5th grader writing assignment result

I see the waves hitting the boat making it rock. I see the wind blowing the coats on the men, the sea gulls making noises and flying over the ship. The men telling merry yarns and helping the other rover row. The sails filling up with wind and moving the ship. The men trying, but can't deny the tide with the waves hitting the boat and all the water flying up. The rovers not knowing where they are going, just sailing wherever the waves are rough. Now I see why God created the waters, for a man who wants to be free.

About the Author:

Serving aboard a ballistic missile submarine and leading the efforts to create a certified stroke program at a midwestern hospital as an RN are his proudest achievements. His name is Rick Palumbo. He's currently retired from the medical field and pursuing his interest in writing through self-publishing. Prior to retiring, he was a registered nurse (RN) for fifteen years and before that an engineer for sixteen years. His first college degree in Paper Science and Engineering was quickly followed with six years in the Navy.

It was the diversity of his background and experiences that led him to select the subject for and write his first book. Over the years he has answered many questions about his submarine experience, coupled with his decision to obtain degrees in engineering and medical fields. This book provides perspective into the seemingly odd combination of papermaking science and submarine experience.

CPSIA information can be obtained
at www.ICGtesting.com
Printed in the USA
BVHW081722120123
656162BV00002B/195